D0065562

# Capillary Electrochromatography

# RSC Chromatography Monographs

Series Editor: Roger M. Smith, *University of Technology, Loughborough, UK*

Advisory Panel: J.C. Berridge, *Sandwich, UK*; G.B. Cox, *Illkirch, France*; I.S. Lurie, *Virginia, USA*; P.J. Schoenmaker, *Eindhoven, The Netherlands*; C.F. Simpson, *London, UK*; G.G. Wallace, *Wollongong, Australia*.

This series is designed for the individual practising chromatographer, providing guidance and advice on a wide range of chromatographic techniques with the emphasis on important practical aspects of the subject.

Supercritical Fluid Chromatography
edited by Roger M. Smith, *University of Technology, Loughborough, UK*

Packed Column SFC
by T.A. Berger, *Berger Instruments, Newark, Delaware, USA*

Chromatographic Integration Methods, Second Edition
by Norman Dyson, *Dyson Instruments Ltd, UK*

Separation of Fullerenes by Liquid Chromatography
edited by K. Jinno, *Toyohashi University of Technology, Japan*

HPLC: A Practical Guide
by Toshihiko Hanai, *Health Research Foundation, Kyoto, Japan*

Applications of Solid Phase Microextraction
edited by Janusz Pawliszyn, *University of Waterloo, Ontario, Canada*

Capillary Electrochromatography
edited by Keith D. Bartle, *University of Leeds, UK*
and Peter Myers, *X-tec Consulting Ltd, UK*

*How to obtain future titles on publication*

A standing order plan is available for this series. A standing order will bring delivery of each new volume upon publication. For further information please contact:

Sales and Customer Care, Royal Society of Chemistry, Thomas Graham House, Science Park, Milton Road, Cambridge CB4 0WF
Telephone: +44(0) 1223 420066

RSC
CHROMATOGRAPHY
MONOGRAPHS

# *Capillary Electrochromatography*

Edited by

**Keith D. Bartle**
*School of Chemistry, University of Leeds, UK*

**Peter Myers**
*X-tec Consulting Ltd, UK*

ROYAL SOCIETY OF CHEMISTRY

ISBN 0-85404-530-9

A catalogue record for this book is available from the British Library

Published by The Royal Society of Chemistry,
Thomas Graham House, Science Park, Milton Road,
Cambridge CB4 0WF, UK

For further information see our web site at www.rsc.org

Typeset by Paston PrePress Ltd, Beccles, Suffolk
Printed by MPG Books Ltd, Bodmin, Cornwall

# *Preface*

More than twenty years ago, new columns were developed for high performance liquid chromatography packed with small-diameter particles. Since then, HPLC has become the preferred analytical separations method for the very large group of compounds which are insufficiently volatile to be analysed in the vapour phase by gas chromatography. HPLC has proved its worth in the analysis of pharmaceuticals and biochemicals, as well as mixtures of environmental interest, and its position is very likely to be secure in the near future. For mixtures requiring very high resolutions, however, HPLC is limited by its reliance on small particles; driving the mobile phase through a packed bed requires high pressures. Reducing the particles below two microns and/or increasing the length of the column increases the pressure requirements above those of existing equipment. Targeted separations in HPLC are usually achieved by the use of specially tailored mobile phases, which allow specific interactions with analytes.

Over the past decade, however, a technique which complements HPLC and overcomes some of its inherent limitations has been developed. In capillary electrochromatography (CEC) the liquid mobile phase is driven through a small-diameter packed column by electroosmotic flow. There is now no pressure drop across the column so that small particles and longer columns can be used. Secondly, electroosmotic flow results in a plug flow profile as opposed to a parabolic flow as derived from pressure driven flow. Both of these advantages should lead to highly efficient columns that can be applied to separate mixtures by 'brute force' efficiency, rather than selectivity. Further driving forces for using CEC are that the small columns allow the analysis of much smaller samples than are used in conventional HPLC, along with the fact that separation by different rates of electromigration of charged analytes can be used to complement the separation of the mixture.

In this book, we have brought together contributions on the theory, practice and application areas of CEC from researchers active in the recent developments.

Keith Bartle, Peter Myers
Leeds, November 2000

# Contents

# Contributors

**Keith D. Bartle,** *School of Chemistry, University of Leeds, Leeds LS2 9JT, UK*
**Maria G. Cikalo,** *School of Chemistry, University of Leeds, Leeds LS2 9JT, UK*
**An Dermaux,** *Department of Organic Chemistry, University of Gent, Krijgslaan 281 (S.4), B-9000 Gent, Belgium*
**Monika M. Dittmann,** *Agilent Technologies, Waldbronn Analytical Division, PO Box 1280, D-76337, Germany*
**Melvin R. Euerby,** *Astra Charnwood, Pharmaceutical and Analytical R & D, Bakewell Road, Loughborough, Leicestershire LE11 5RH, UK*
**Nicola C. Gillott,** *School of Pharmaceutical Sciences, University of Nottingham, Nottingham NG7 2RD, UK*
**D.B. Gordon,** *Department of Biological Sciences, Manchester Metropolitan University, Chester Street, Manchester M1 5GD, UK*
**G.A. Lord,** *MRC Toxicology Unit, Hodgkin Building, University of Leicester, Lancaster Road, Leicester LE1 9HN, UK*
**Peter Myers,** *X-tec Consulting Ltd, Woodlea, Bromborough, Wirral CH62 6DL, UK*
**Vincent T. Remcho,** *Department of Chemistry, Oregon State University, Corvallis, Oregon 97331, USA*
**Mark M. Robson,** *School of Chemistry, University of Leeds, Leeds LS2 9JT, UK*
**Gerard P. Rozing,** *Agilent Technologies, Waldbronn Analytical Division, PO Box 1280, D-76337, Germany*
**Pat Sandra,** *Department of Organic Chemistry, University of Gent, Krijgslaan 281 (S.4), B-9000 Gent, Belgium*
**Norman W. Smith,** *Zeneca/SmithKline Beecham Centre for Analytical Science, Department of Chemistry, Imperial College of Science, Technology and Medicine, South Kensington, London SW7 2AY, UK*
**Patrick T. Vallano,** *Department of Chemistry, Oregon State University, Corvallis, Oregon 97331, USA*

# Symbols and Abbreviations

## Symbols

| | |
|---|---|
| $A$ | eddy diffusion coefficient |
| $A_p$ | free cross sectional area in packed section of capillary |
| $A_{Ot}$ | free cross sectional area in open section of capillary |
| $a$ | distance from tube centre |
| $B$ | longitudinal diffusion plate height coefficient |
| $C$ | mass transfer plate height coefficient |
| $c$ | electrolyte concentration |
| $\delta$ | double layer thickness |
| $D_m$ | solute diffusion coefficient in mobile phase |
| $d_{ch}$ | channel diameter |
| $d_p$ | particle diameter |
| $E$ | electric field strength |
| $E_{appl}$ | applied field strength |
| $E_{eff}$ | effective field strength |
| $\epsilon_0$ | permittivity of vacuum |
| $\epsilon_r$ | dielectric constant |
| $\varepsilon$ | porosity |
| $F$ | Faraday's constant |
| $h$ | reduced plate height |
| $H$ | plate height |
| $H_{open,FP}$ | plate height due to flow profile distortion in open section |
| $H_{TH}$ | plate height due to thermal effects |
| $\eta$ | solution viscosity |
| $I$ | ionic strength |
| $I_0$ | zero order Bessel function of the first kind |
| $k$ | retention factor |
| $\kappa$ | reciprocal of double layer thickness, thermal conductivity of mobile phase |
| $L$ | column length |
| $L_{open}$ | length of open section |
| $L_p$ | length of packed section |

| $L_{tot}$ | total column length |
| $\lambda$ | molar conductivity of electrolyte |
| $Q$ | heat generated per unit volume of electrolyte |
| $P_i$ | intersegmental pressure |
| $P_0$ | pressure at capillary inlet and outlet |
| $R$ | gas constant |
| $r$ | capillary tube radius |
| $\sigma$ | charge density at shear surface |
| $\sigma_l^2$ | length variance of solute band |
| $\sigma_p$ | conductivity in packed section |
| $\sigma_{ot}$ | conductivity in open section |
| $T$ | temperature (Kelvin) |
| $\Delta T$ | temperature difference between tube centre and inner wall |
| $\mu_{eo}$ | electroosmotic flow velocity |
| $\upsilon$ | reduced velocity |
| $v$ | volumetric flow rate |
| $\psi$ | potential |
| $z$ | electrolyte valence |
| $\zeta$ | zeta potential |

# Abbreviations

| AMPS | 2-acrylamido-2-methyl-1-propanesulfonic acid |
| APCI | atmospheric pressure chemical ionization |
| CAD | collisionally assisted dissociation |
| CE | capillary electrophoresis |
| CEC | capillary electrochromatography |
| CF | continuous flow |
| CF-FAB | continuous flow fast atom bombardment |
| CN | carbon number |
| CZE | capillary zone electrophoresis |
| DAD | diode array detection |
| DMSO | dimethyl sulfoxide |
| ELSD | evaporative light-scattering detection |
| EOF | electroosmotic flow |
| ESI | electrospray ionization |
| ESMS | electrospray mass spectroscopy |
| FAB | fast atom bombardment |
| FAME | fatty acid methyl ester |
| FAPE | fatty acid phenacyl ester |
| FFA | free fatty acid |
| GC | gas chromatography |
| HETP | height equivalent to a theoretical plate |
| HPLC | high-performance liquid chromatography |
| MALDI | matrix-assisted laser desorption ionisation |
| MeCN | acetonitrile |

| | |
|---|---|
| MEKC | micellar electrokinetic chromatography |
| MIP | molecular imprint polymer |
| MRM | multiple reaction monitoring |
| MS | mass spectrometry |
| NSAIDS | non-steroidal anti-inflammatory drugs |
| ODS | octadecylsilane |
| OT | open tubular |
| PEC | pseudo-electrochromatography |
| PN | partition number |
| PTFE | poly(tetrafluoroethylene) |
| RP-pSFC | reversed-phase packed column supercritical fluid chromatography |
| RSD | relative standard deviation |
| SAX | strong anion exchange |
| SCX | strong cation exchange |
| SFC | supercritical fluid chromatography |
| SI-pSFC | silver ion packed column supercritical fluid chromatography |
| SIR | selected ion recording |
| TEA | triethylamine |
| TEOS | tetraethoxysilane |
| TOF | time-of-flight |
| TRIS | tris(hydroxymethyl)methylamine |
| UV | ultraviolet |
| UV–Vis | ultraviolet–visible |

CHAPTER 1

# An Introduction to Capillary Electrochromatography

KEITH D. BARTLE, MARIA G. CIKALO AND
MARK M. ROBSON

## 1 What is Capillary Electrochromatography?

Capillary electrochromatography (CEC) is a recently developed variant of
high-performance liquid chromatography (HPLC) in which the flow of mobile
phase is driven through the column by an electric field, a phenomenon known
as electroosmosis, rather than by applied pressure. This electroosmotic flow
(EOF) is generated by applying a large voltage across the column; positive
ions of the added electrolyte accumulate in the electrical double layer of
particles of column packing, move towards the cathode and drag the liquid
mobile phase with them. As in capillary electrophoresis (CE) and micellar
electrokinetic chromatography (MEKC), small diameter (typically 50–100 $\mu$m)
columns with favourable surface area-to-volume ratio are employed to
minimise thermal gradients from ohmic heating, which can have an adverse
effect on band widths. CEC differs crucially from CE and MEKC, however, in
that the separating principle is partition between the liquid and solid phases
(Table 1.1).

Avoiding the use of pressure results in a number of important advantages for
CEC over conventional HPLC. Firstly, the pressure-driven flow rate through a
packed bed depends directly on the square of the particle diameter and inversely
on column length; for practical pressures, generally used particle diameters are
seldom less than 3 $\mu$m, with column lengths restricted to approximately 25 cm.
By contrast the electrically driven flow rate is independent of particle diameter
and column length so that, in principle, smaller particles and longer columns
can be used. If follows that considerably higher efficiencies can be generated in
CEC than in HPLC. A second consequence of employing electrodrive is that the
plug-like flow-velocity profile in EOF reduces dispersion of the band of solute as
it passes through the column, further increasing column efficiency. The
combined effect of reduced particle diameter, increased column length and

**Table 1.1** *Comparison of electrically driven separation methods*

|                             | CE                                               | MEKC                                                                                            | CEC                                                       |
| --------------------------- | ------------------------------------------------ | ----------------------------------------------------------------------------------------------- | -------------------------------------------------------- |
| Separation principle        | Different mobilities of ions in electric field   | Partition between bulk solution and micelle moving in opposite direction to analyte             | Partition between solid stationary phase and mobile phase |
| Typical column diameter/$\mu$m | 50–100                                        | 50–100                                                                                          | 50–100                                                   |
| Stationary phase            | None                                             | None                                                                                            | Silica particles with bonded groups                      |
| Mobile phase                | Electrolyte solution                             | Electrolyte solution                                                                            | Electrolyte solution                                     |
| Sample type                 | Charged species                                  | Neutrals                                                                                        | Neutrals and charged species                             |

plug flow leads to CEC efficiencies of typically 200 000 plates per metre, and substantially improved resolution.

Voltages up to 30 kV are applied to generate the electric field usually for solutions of 1–50 mM buffers in aqueous reversed-phase mobile phases, although non-aqueous CEC has also been carried out. The dependence of EOF rate on solvent dielectric constant has been confirmed, but the electrical potential (the zeta potential) of the boundary between the fixed and diffuse layers (the double layer; see pages 43–45 for further discussion) of positive ions at the stationary phase wall (Figure 1.1) is less well understood. The conclusion of an early theoretical study which suggested that flat EOF profiles in a capillary of diameter $d$ would result if $d$ were considerably greater than the double layer thickness, $\delta$, has been confirmed by experiment; for channels between particles, however, the influence of $\delta$ is less clear. Current indications are that it should be possible to use monodisperse particles with diameters down to 0.5 $\mu$m. Pore sizes of commonly used HPLC particles are too small to give rise to EOF, but larger pore packings show promise. Although CEC has been demonstrated for stationary phases bonded to the walls of open tubes, and in sol–gel derived phases, most work has been carried out on columns packed with HPLC stationary phases; a new generation of packings custom-synthesized for CEC is, however, now beginning to make an impact.

## 2   History of CEC

Strain[1] first reported the use of the EOF in chromatography; he recognized the difference between electrophoresis and electrochromatography on the one hand, and partitioning of analytes between a mobile and a stationary phase on the

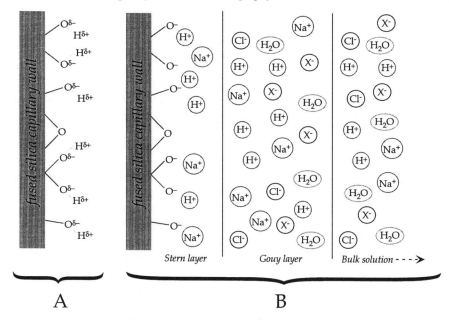

**Figure 1.1** (A) *Representation of the surface of fused silica tubing.* (B) *Formation of an electrical double layer near the surface of fused silica tubing* (Courtesy of V.T. Remcho and P.T. Vallano)

other. Electrodrive (electrophoretic mobility and electroosmosis) was used to move the analytes through a separation medium, so that the importance of the EOF was recognized in electrochromatography. Early work in electrical chromatography was either in relatively large diameter columns ($>1$ mm)[2] or in thin layers, which were used to analyse neutral, basic and acidic molecules by electromigration through a paper matrix.[3]

The separation of polysaccharides using electrodrive through a colloidal membrane is probably the first reported use of EOF to drive a mobile phase through a stationary phase.[4] For thin layers, Kowalczyk[5] quantified the EOF velocity while Hybarger *et al.*[6] proposed an annular bed system for preparative separations. Cylindrical columns packed with Sephadex were used by a number of groups for protein separation.[7,8] Gel columns were employed in the separation of high-molecular-weight ($>500$ kDa) compounds in experiments in which counteracting electrophoretic and hydrodynamic forces were used.[9]

The originators of CEC were, however, the Pretorius group,[10] who reported that if the EOF were used to drive the mobile phase flow, as opposed to the hydrodynamic flow in conventional liquid chromatography, the plate height was reduced. They also pointed out the absence of pressure drop across the column if the EOF were used. Significant progress in CEC began in the 1980s; Jorgenson and Lukacs[11] demonstrated the use of electroosmosis in capillaries and showed the possibilities for low reduced plate heights. Tsuda then showed that CEC was possible in coated open tubular columns and recognized the factors that control the EOF as well as the importance of practical effects, such as bubble formation, in packed columns.[12]

The recent resurgence of CEC dates from the detailed theoretical analysis of Knox and Grant, published in 1987,[13] followed by practical demonstrations by the same group in 1991.[14] Both slurry-packed and draw-packed capillaries were used in a detailed study of factors affecting the EOF. Particle sizes down to 1.5 $\mu$m were used, and reduced plate heights near unity were demonstrated in 30–200 $\mu$m i.d. columns up to 1 m long. The important observation was made that columns driven electrically show higher efficiencies than the same column with pressure drive (Figure 1.2).

The potential of CEC in the analysis of mixtures relevant to the pharmaceutical industry was realized by Smith and Evans in 1994.[15] The capabilities of CEC were demonstrated in high-resolution chromatograms of drug compounds (*e.g.* Figure 1.3) on a reversed-phase $C_{18}$ stationary phase. The same group than went on to show how especially strong cation-exchanger stationary phases allowed analyte focusing and very narrow peaks for basic drugs.[16] Some landmarks in the history of CEC are listed in Table 1.2.

Since the re-emergence of CEC, much work has centred on the establishment of reliable column technology (Chapter 3), on identifying suitable mobile phase buffers[17] and on investigating the theory and mechanism of CEC (Chapter 4). Since 1996, however, there has been a very rapid increase in the number of publications and reviews[18–20] relating to CEC. The numbers of reports of CEC separations of compounds from the environment, and of biomolecules, are growing rapidly,[18,19] while the complementary relation of CEC separations to generally used HPLC methods has led to much activity in the analysis of

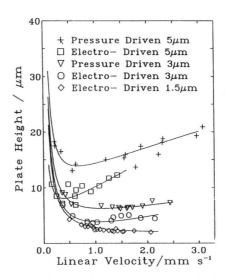

**Figure 1.2** *Comparison of plate heights for pressure- and electro-driven columns for different particle sizes*
(Reproduced with permission from ref. 14)

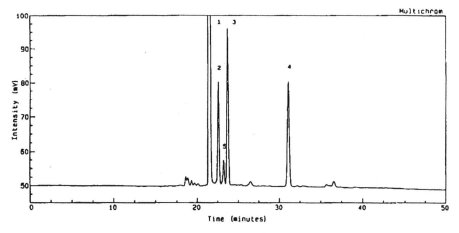

**Figure 1.3** *Separation by CEC of fluticasone propionate* (1) *and related impurities* (2–5)
(Reproduced with permission from ref. 15)

**Table 1.2** *Landmarks in CEC*

| Event | Year | Reference |
|---|---|---|
| First report of use of EOF in chromatography | 1939 | 1 |
| Separation of polysaccharides using EOF through colloidal membrane | 1954 | 4 |
| Use of EOF in column chromatography | 1974 | 10 |
| Electroosmosis in capillaries | 1981 | 11 |
| CEC in open tubular columns | 1987 | 12 |
| Theory of CEC and technique development | 1987, 1991 | 13,14 |
| Analysis of pharmaceutical compounds by CEC | 1994 | 15 |

pharmaceuticals by CEC (Chapter 7). The extremely low flow rates in CEC
($< 1$ $\mu$L min$^{-1}$) help make coupling to mass spectrometry an especially attractive possibility (Chapter 6), first demonstrated in 1991.[21]

# 3 Electroosmosis

Electroosmosis is best described as the movement of liquid relative to a
stationary charged surface under an applied electric field. Substances tend to
acquire a surface charge as a result of ionization of the surface and/or by
interaction with ionic species. In a fused silica capillary, the ionization of silanol
groups gives rise to a negatively charged surface, which affects the distribution
of nearby ions in solution. Ions of opposite charge (counterions) are attracted to

the surface to maintain the charge balance whilst ions of like charge (co-ions) are repelled. The double layer of electric charge thus formed (see Figure 1.1) is generally explained by a revised version of the Gouy–Chapman model.[22] Essentially the counterions are arranged in two layers, fixed and diffuse, with a surface of shear at just beyond the interface. The voltage drop between the wall and this surface of shear is known as the zeta potential, $\zeta$. In the diffuse layer, the potential falls exponentially to zero, and the distance over which it falls by $e^{-1}$ is known as the double layer thickness, $\delta$. When the voltage is applied, the solvated cations in the diffuse layer migrate towards the cathode, dragging the solvent molecules along with them.

The linear velocity of the EOF, $u_{eo}$, is described[23] by the Smoluchowski equation

$$u_{eo} = \frac{\varepsilon_0 \varepsilon_r \zeta E}{\eta} \tag{1.1}$$

This shows how the EOF is governed by $\zeta$, the permittivity and viscosity of the mobile phase, $\varepsilon_r$ and $\eta$, and the electric field strength, $E$.

The flow profile is assumed to be near-plug-like since essentially it originates from the capillary wall, but in reality it depends on the capillary internal diameter, $d$, and $\delta$. Theoretical studies by Rice and Whitehead[24] proposed that $u_{oe}$ is only independent of the capillary diameter when $d \gg \delta$. As $d$ approaches $\delta$, double layer overlap occurs with a simultaneous reduction in flow velocity, until finally a parabolic flow profile is obtained when $d$ and $\delta$ are similar. It has been proposed that the EOF velocity is acceptable when $d \geq 10\delta$.[13] The use of microscope optics to image flow profiles in narrow capillaries has produced conflicting results. Whilst the plug flow profile predicted from theory has been observed by Taylor and Yeung,[25] Tsuda *et al.*[26] found a higher EOF at the capillary wall than at the centre. The importance of the EOF profile in CEC necessitates further research in this area.

From equation (1.1) it can be seen that neither the diameter of the stationary phase particle ($d_p$) nor the column length ($L$) affect the mobile phase velocity. This is in contrast with pressure-driven flow velocity $\bar{v}$ which is described by the Kozeny–Carman equation[27]

$$\bar{v} = \frac{\varepsilon^2}{180(1-\varepsilon)^2} \frac{d_p^2}{\eta} \frac{\Delta p}{L} \tag{1.2}$$

where $\varepsilon$ is the porosity (approximately 0.4 for a randomly packed column) and $\Delta p$ is the pressure drop.

In packed CEC, both the capillary wall and the column packing carry surface charges that are capable of supporting EOF. To date, most of the work carried out suggests[28] that the column packing is responsible for the generation of EOF; there is a greater number of free silanol groups present since the solid packing has a far larger surface area compared to that of the internal silica wall. If the column is assumed to consist of a closely-packed array of non-porous spherical particles, then the EOF arises from the channels between the particles. The average interparticle channel is estimated to be one-quarter to one-fifth the particle diameter.[29] It has been suggested that the EOF velocity in a CEC column is most likely to be reduced compared to that in an open tube, on account of the tortuosity and porosity of the packed bed. Although there does not appear to be any adverse effect as a result of packing irregularities[11] further investigations have been made on packing structure using electrical conductivity measurements.[30,31]

Using capillary columns comprising a packed and an open section, it was possible to calculate the voltage drop in each section, and thus to determine the individual contributions to the total EOF.[31] Measurements made at different pH for columns packed to different lengths with reversed-phase octadecyl silica (ODS) and strong cation-exchange (SCX) particles showed that: at extremes of pH the electric field strength is far greater in the packed section; at intermediate pH the field strengths are similar in both sections; at pH 7.5 the length of the packed section has little impact on the total conductance of the column, but the conductance of the open section does scale approximately with its length. At pH 7.5 the contributions of the open and packed sections are roughly equal when the capillary is half packed, but the same effect at pH 10.5 is only apparent when the capillary is about one-quarter packed. This study[31] confirmed that, except at low pH, the EOF in a packed column is generally less than that in an open capillary, and that the greatest influence of length of packed bed on $u_{eo}$ is observed at low or high pH.

# 4   Classification of CEC

Separation in CE is a consequence of differential migration of charged and neutral species (Figure 1.4). CEC may be compared with CE and classified in the hierarchy of separation methods employing liquids by the unified description proposed by Rathore and Horvath.[32] The differential migration of solute bands can be divided into components that are separative (selective interactions with a stationary phase, or differences in electrophoretic migration velocities), and components that are non-separative (migration not contributing directly to separation). The concept of 'virtual migration lengths'[33] then allows the CEC retention factor, $k_c$ to be defined. Noting that the separative (HPLC) retention factor, $k$, is given by

$$k = \frac{t_R - t_0}{t_0} \tag{1.3}$$

and that the CE velocity factor, $k_e$, is, correspondingly

$$k_e = \frac{t_{eo} - t_m}{t_m} \tag{1.4}$$

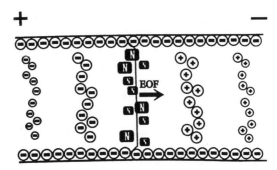

**Figure 1.4**  *Differential solute migration in a column under a voltage gradient*

where $t_R$ and $t_m$ are elution and migration times respectively, $t_0$ is the retention time of an unretained marker, and $t_{eo}$ is the retention time of a solute moved only by the EOF, then,

$$k_c = k + kk_e + k_e \qquad (1.5)$$

The product $kk_e$ is the consequence of simultaneous chromatography and electrophoresis. If $k_e = 0$ then $k_c$ simply equals $k$, and only HPLC operates. Correspondingly, if $k = 0$, then $k_c = k_e$ and the only process is CE.

In HPLC, and in CEC of neutral species (case A in Figure 1.5), the solute and the mobile phase move in the same direction, and the sample components emerge in order of retardation by the stationary phase. However, if the solute is charged, there are three operational modes depending on the direction and magnitude of electrophoretic migration with respect to the direction of the EOF (Figure 1.5):

  B.  co-directional, where the migration velocities of charged species are always greater than that of the EOF marker. The components emerge before the EOF marker;
  C.  counter-directional, where the EOF velocity is greater than the electrophoretic velocity of a charged component, which emerges after the EOF marker;
  D.  counter-directional, where the EOF velocity is less than the electrophoretic velocity, and detection of a charged component is only possible with instrument polarity reversed. In this case, the EOF marker is not detected.

The order of emergence of different mixture components relative to each other depends on the combination of their different retardation and migration velocities, in accordance with equation (1.5).

CEC offers the substantial advantage over MEKC of a time window for separation which is, in principle, unlimited. In MEKC, however, all electrically neutral compounds have migration times between $t_0$ and $t_{mc}$, that of the micelle.

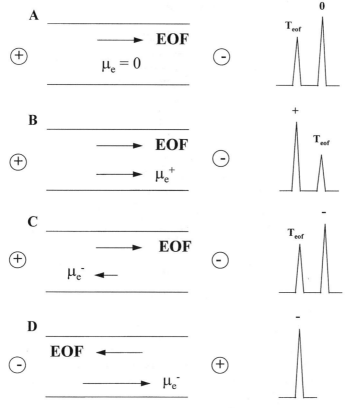

**Figure 1.5** *CEC of charged species. $T_{eof}$ denotes retention of EOF marker*

A hybrid of MEKC and CEC has been proposed by Knox.[34] In colloidal sol electrochromatography, a colloidal sol is used as the moving fluid. If the colloidal particles are charged they move relative to the eluent, and partition between two phases occurs resulting in separation.

## 5 Band-broadening in CEC

In column chromatography, a number of processes bring about the broadening of solute bands: (a) eddy diffusion, originating from the variety of flow paths through the packed bed; (b) axial molecular diffusion; (c) resistance to mass transfer in the mobile and stationary phases and (d) system effects, such as those arising from dead volumes. The smaller theoretical plate heights ($H$) and hence greater plate numbers $N$ ($= L/H$) in CEC in comparison with conventional pressure-driven HPLC arises from reduced contributions to $H$ from factors (a) and (c) above.

Figure 1.6 illustrates the differences in flow velocity profiles in the packed bed. Clearly, the plug flow profile of CEC substantially reduces the eddy diffusion (or

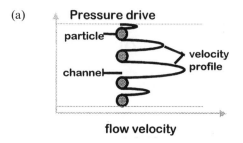

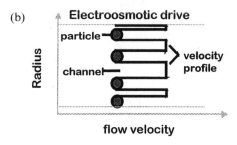

**Figure 1.6** *Flow velocity profiles in a packed bed with* (a) *pressure drive;* (b) *electroosmotic drive*

multipath) term in comparison with the parabolic flow profile of HPLC. Since this term is also proportional to the column particle diameter, the use of smaller particles should further reduce the contribution to $H$; the contribution from slow mass transfer in the mobile phase, $C_m\bar{u}$ where $\bar{u}$ is average linear mobile phase velocity, is proportional to $d_p^2$.

The use of EOF to drive the mobile phase flow gives rise (Figure 1.6) to a plug flow profile in the channels between the particles. This also reduces the mass transfer contribution to $H$ by a factor which can be estimated by considering such a channel as an open tube, of diameter $d_c$, for which $C_m$ is given by the Golay equation:[35]

$$C_m = \frac{f(k)d_c^2}{D_m} \tag{1.6}$$

where $D_m$ is the diffusion coefficient of the solute in the mobile phase. For parabolic flow, as in HPLC

$$f(k) = \frac{1 + 6k + 11k^2}{96(1 + k)^2} \tag{1.7}$$

but for plug flow, in CEC[36]

$$f(k) = \frac{k^2}{16(1 + k)^2} \tag{1.8}$$

It follows that, for a given $k$, and the same $d_c$ and $D_m$, the contribution to $H$ from this source for CEC is about half that in capillary HPLC.

The model developed by Horvath and Lin[37] to describe band broadening in HPLC was applied to CEC by Dittman *et al.*[38] An improvement in efficiency over HPLC of approximately 100% was predicted, so that minimum reduced plate heights $H_r$ ($= H/d_p$) near unity are predicted; this has generally been borne out by experiment for CEC for columns packed with particles with $d_p \geq 3$ $\mu$m (see for example Figure 1.7). The small increase in $H$ with increasing $\bar{u}_{opt}$, the mobile phase velocity at the minimum value of $H$, suggests that efficient CEC should be possible at high $\bar{u}$, thus shortening analysis times. Knox and Grant[13] calculated that the minimum value of $d_p$ which can be used without affecting the EOF velocity is 0.5 $\mu$m, but, perhaps because columns are more difficult to pack with small particles (Chapter 3) their reported applications have been comparatively few. Lüdtke and co-workers have packed 0.5 $\mu$m silica beads, but observed minimum reduced plate heights of $\sim 3.5$–$4.00$.[40]

In CEC an extra factor that may decrease column efficiency compared to HPLC is ohmic heating in the capillary. Knox and Grant showed[13] that the column diameter must be greatly reduced for electrodrive in comparison with pressure drive to avoid temperature gradients, which contribute to band broadening. Boughtflower *et al.*[17] also recommended the use of low conductivity zwitterionic buffers such as tris(hydroxymethyl)methylamine (TRIS) to reduce self-heating problems. In an experimental study[41] graphs of $H$, against $\bar{u}$ show increased efficiency for 50 and 75 $\mu$m i.d. columns in comparison with 100 $\mu$m i.d. for the same packing material.

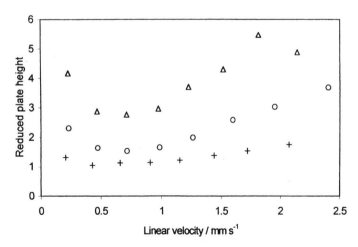

**Figure 1.7** *Van Deemter plots[39] for ethylparaben on 3 ($\triangle$), 5 ($\bigcirc$) and 10 ($+$) $\mu$m ODS1. Applied voltage: 2.5–25 kV. Electrolyte: phosphate pH 7.5 containing 70% v/v acetonitrile*
(Reproduced with permission from ref. 39)

# 6   Effect of Operating Variables in CEC

From the theoretical relationship (equation 1.1) between EOF linear velocity and electric field strength, $u_{eo}$ should be directly proportional to $E$, or to the applied voltage, $V$, if the capillary length is kept constant. While deviations from linearity in graphs of $u_{eo}$ against $V$ have been observed at high field strengths, because of ohmic heating, which reduces viscosity, equation (1.1) is obeyed up to 20 kV for mixed solvent systems (see Figure 1.8).[39]

The values of $u_{eo}$ for 3, 5 and 10 $\mu m$ particles are similar, and since any differences are observed at higher electric field strengths (Figure 1.8) it is probable that variations in $u_{oe}$ are related to the effects of ohmic heating. One may surmise that the dissipation of heat depends on packing efficiency, itself dependent on $d_p$.

Since $u_{eo}$ is proportional to the zeta potential (equation 1.1), which is related to the capillary surface charge, $\sigma$, and the double-layer thickness, $\delta$, via equation 1.9[22]

$$\zeta = \frac{\sigma\delta}{\varepsilon_0\varepsilon_r} \tag{1.9}$$

it follows that factors that influence $\sigma$ and $\delta$ will affect $u_{eo}$, in particular the pH and ionic strength ($I$) of the mobile phase. It has been shown[39] that $u_{eo}$ decreases with pH for a variety of column packings (see Figure 1.9) because of reduced ionization of residual silanol groups at low pH. However, it is important to note

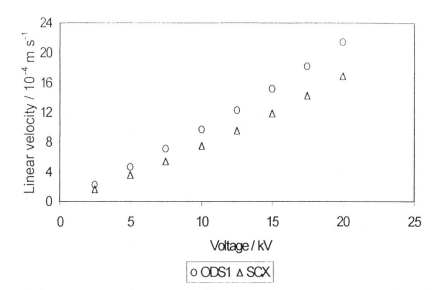

**Figure 1.8**   *Effect of applied voltage on EOF linear velocity for ODS1 ($\bigcirc$) and SCX ($\triangle$) packings*
(Reproduced with permission from ref. 39)

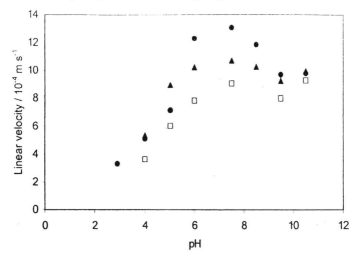

**Figure 1.9** *Dependence of EOF linear velocity on pH in CEC (▲, ODSI and ●, silica) and CE (□). Electrolytes of varying pH of approximately 10 mM ionic strength containing 70% v/v acetonitrile*
(Reproduced with permission from ref. 39)

that results of this kind are often reported with the use of electrolytes of variable $I$, making comparisons difficult.

A decrease in $I$ will cause an increase in the double layer thickness according to the relation[42]

$$\delta = \left(\frac{\varepsilon_0 \varepsilon_r RT}{2F^2 I}\right)^{1/2} \tag{1.10}$$

where $R$, $T$ and $F$ are respectively the universal gas constant, temperature and Faraday constant.

The effect of ionic strength has been studied[39] over the range 1–20 mM phosphate in electrolytes containing 70% (v/v) acetonitrile. Short columns packed with 3 $\mu$m ODSI were utilized and it was assumed the pH was 7.5 throughout. As the ionic strength is reduced the EOF is expected to increase unless double layer overlap occurs. Since the packing material has a pore size of 8 nm, double layer overlap in the pores is expected to occur at ionic strengths of 2.5 mM and below where the double layer is at least 4 nm thick (see Table 1.3).

**Table 1.3** *Calculated double-layer thickness (nm) in aqueous and acetonitrile/water (70:30) systems for varying ionic strength solutions*

| $I$/mol m$^{-3}$ | 1.0 | 2.5 | 5.0 | 10.0 | 15.0 | 20.0 |
|---|---|---|---|---|---|---|
| $\delta$(aqueous)/nm | 9.6 | 6.1 | 4.3 | 3.0 | 2.5 | 2.2 |
| $\delta$(70:30)/nm | 7.6 | 4.8 | 3.4 | 2.4 | 2.0 | 1.7 |

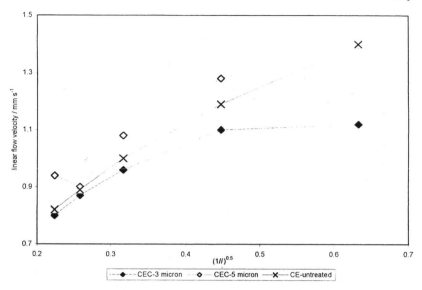

**Figure 1.10**   *Comparison of the effect of ionic strength in CEC and CE. Data from ref. 39*

Figure 1.10 illustrates that the increases in EOF for CE and CEC are generally similar at the higher ionic strengths, but below 5 mM the EOF in CEC starts to level off and drop slightly. By contrast, in CE the EOF continues to rise. This marked change in the EOF at lower ionic strengths would suggest that double layer overlap was occurring within the pores of the packing material.

Care is also necessary in assuming that variation in $I$ does not affect investigations of the influence of the type and proportion of organic solvent in the mobile phase. CE experiments have been performed[39] using a phosphate electrolyte, which in water would give an ionic strength of 10 mM and a pH of 7.5, containing between 0 and 90% (v/v) of different organic modifiers: acetone, acetonitrile, methanol and propan-2-ol. It was assumed that the buffer component made no contribution to $\varepsilon_r$ and $\eta$, and the ratios of $\varepsilon_r$ to $\eta$ were predicted for 0, 50 and 100% organic (Table 1.4). From these values and equation (1.1), an

**Table 1.4**   *Predicted ratio of permittivity to viscosity, $\varepsilon_r/\eta$ $(cP^{-1})$ for solvent/ water mixtures at 25 °C*

|  | *Organic content (%)* | | |
|---|---|---|---|
| *Solvent* | 0 | 50 | 100 |
| Acetone | 88 | 23 | 68 |
| Acetonitrile | 88 | 75 | 105 |
| Methanol | 88 | 37 | 60 |
| Propan-2-ol | 88 | 15 | 7 |

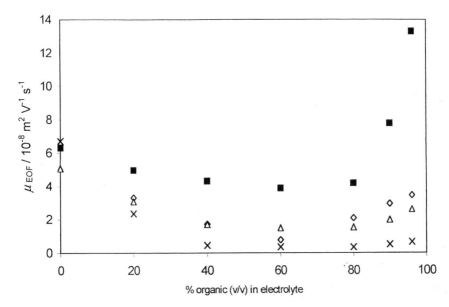

**Figure 1.11**  *Effect of organic content of electrolyte on EOF mobility (■, acetonitrile; ◇, acetone; △, methanol; ×, propan-2-ol)*

indication of the expected EOF behaviour with increasing modifier content can be obtained. For the electrolyte containing propan-2-ol a general decrease in EOF would be expected as the proportion is increased, whilst for the other modifiers a minimum in the EOF should occur. Experimental findings (Figure 1.11) are in good agreement, with acetone, acetonitrile and methanol all showing an EOF minimum around 50–70% organic, although the increase observed when using even higher acetonitrile content was far greater than anticipated. Wright *et al.*[43] have also witnessed this behaviour for acetonitrile/water systems without supporting electrolyte; they have suggested that it can be explained by changes in solvent polarity and hydrogen-bond donor ability. However, without the phosphate buffer, it was found[39] that mixtures containing above 80% acetonitrile were unable to support EOF. From the above considerations it follows that acetonitrile is most commonly selected as the organic solvent since it gives the highest EOF, the current is low, and there is a comparatively stable EOF region (60–80% acetonitrile) within which the effect of solvent evaporation should be minimal. The use of high acetonitrile contents resulted in sufficient flow for the elution of acid compounds in ion-suppressed mode at pH 2.5.

The influence of the percentage of acetonitrile in the mobile phase on $k$ has been investigated.[44,45] Since (Figure 1.12) ln $k$ is linearly related to percentage acetonitrile content of the mobile phase for constant buffer volume (difference made up with water), it follows that well-established theories used in HPLC method developments are equally applicable to the separation of neutral compounds in CEC. The principle of isoeluotropy has also been tested[44,45] and

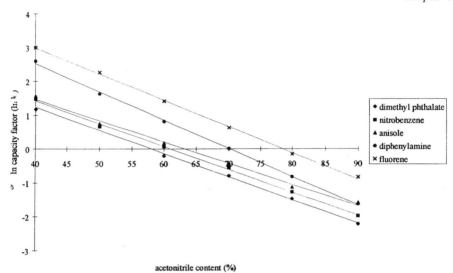

**Figure 1.12**   *Effect of acetonitrile content of the mobile phase on the logarithm of the capacity factor,* k, *on ODS1 stationary phase*[45]

found to operate in reversed-phase CEC. Acetonitrile-based and methanol-based mobile phase compositions of equivalent eluotropic strength gave very nearly equal retention factors (Figure 1.13), although of course, retention times in methanol/water were much longer because of reduced values of the ratio $\varepsilon_r/\eta$.

Increasing the column temperature in CEC increases the EOF for a given voltage and reduces retention times since the mobile phase viscosity falls. As in HPLC, van't Hoff plots of ln $k$ *versus* $1/T$ are linear,[46] but may have different

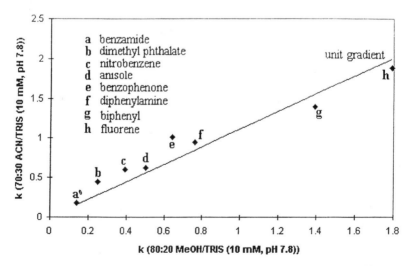

**Figure 1.13**   *CEC retention factors for mobile phases isoelutropic in HPLC*[45]

slopes for different compounds, so that $k$ changes differently with temperature. It follows that selectivity may be tuned by changing the column temperature. Figure 1.14 shows the influence of temperature on the chromatogram of a mixture of diuretics;[41] the coalescence and subsequent 'cross-over' of the peaks of chlorthalidone and hydroflumethiazide is noteworthy.

# 7 Application Areas of CEC

In the past few years the number of published applications of CEC has increased rapidly, and this volume reports developments in all the principal growth areas; pharmaceuticals (Ch. 7), natural products (Ch. 8) and chiral compounds.[19,47]

It is pertinent, however, to examine the potential of CEC, and how it may develop. Since reduced plate heights in CEC can be as low as 1, readily available columns packed with 3 $\mu$m particles can deliver efficiencies of $2 \times 10^5$ theoretical plates m$^{-1}$. Since it should be possible to reduce particle sizes to 0.5 $\mu$m before plug flow is lost, in theory efficiencies as high as $10^6$ plates m$^{-1}$ can be obtained, an improvement of an order of magnitude over HPLC. Moreover, there are no pressure limits on the length of the packed tube; the limit is on the maximum voltage that can be applied, typically up to 30 kV for commercial systems, with reports of experimental systems operating beyond 90 kV.[48]

With such high efficiencies, CEC should deliver real advantages in terms of the column peak capacity, $P$, the number of peaks which can be separated in a chromatogram between realistic capacity factor limits.[49]

$$P = 1 + \frac{N^{1/2}}{4}\ln(1+k) \qquad (1.11)$$

Even though this may represent an overestimate in applications to real mixtures, $P$ is still a useful measure of the resolving power in comparisons of separative methods. Table 1.5 compares peak capacities available in capillary gas chromatography (GC) and HPLC with those theoretically possible in CEC. Clearly, very substantial improvements over HPLC are possible with columns packed with 3 $\mu$m particles, and more especially if the full capability of 1.5 $\mu$m

**Table 1.5** *Peak capacities available in GC, HPLC and CEF*

| Technique | Column length | Theoretical plates | Peak capacity |
|---|---|---|---|
| GC | 50 m | 200 000 | 260 |
| HPLC | 25 cm | 25 000 | 90 |
| CEC | 25 cm (3 $\mu$m particles) | 60 000 | 140 |
| | 50 cm (3 $\mu$m particles) | 120 000 | 200 |
| | 50 cm (1.5 $\mu$m particles) | 200 000 | 260 |

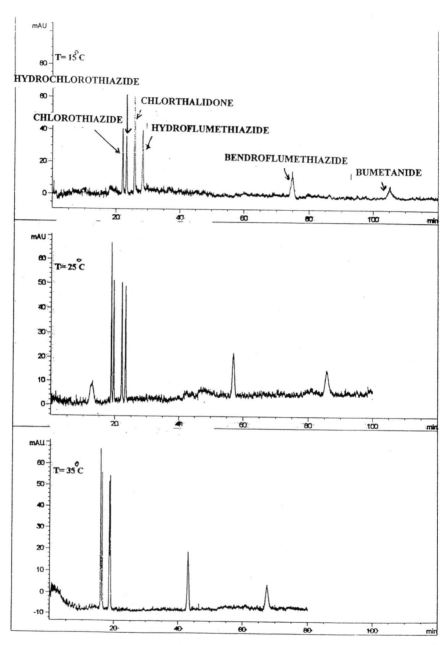

**Figure 1.14**  *Influence of temperature on the chromatography of a mixture of diuretics*[41]

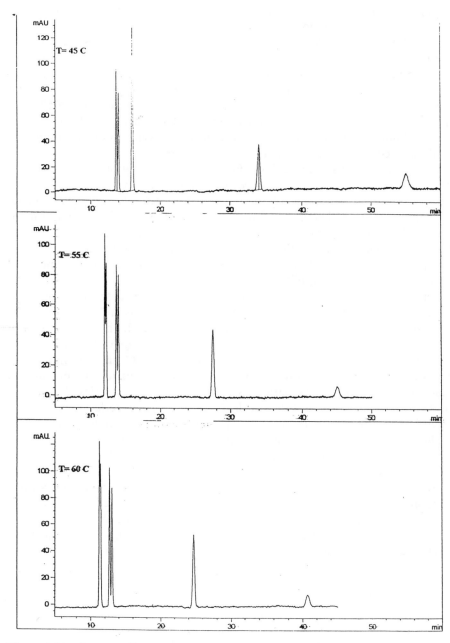

**Figure 1.14** *(continued)*

particle columns is realised. Thus for $N = 2 \times 10^5$ and $k = 10$, $P$ is $>250$ taking CEC into a resolution domain similar to that possible in GC, and which conventional HPLC cannot approach.

Most HPLC separations are currently achieved on the basis of selectivity, but the higher plate numbers of CEC may offer substantial advantages for very complex mixtures of, for example, biological compounds. Thus, CEC has been shown to be a promising technique for the separation of protein digests.[50] Of course, such mixtures can now be separated with high resolution by CE, but there is probably an analogy with the progress of GC, where the advent of fused silica column technology offered the resolution necessary to make routine the analysis of complex fuel and environmental mixtures.[51]

A second major driving force for CEC is the miniaturization of HPLC, which has come about because of the necessity of analysing picomole amounts of substances available in small volumes of body fluids or in the products of single-bead combinatorial chemistry. In fact, it is comparisons of CEC with micro-HPLC that are most meaningful and the necessary development for micro-HPLC of robust, easily installed columns with reliable injection procedures, pumping and gradient elution methods, parallels the practical requirements of CEC for future routine application.

The great test of CEC will be whether regulatory authorities will accept its use in analysis, especially of pharmaceuticals. Here, precision, accuracy, trace analysis and repeatability are vital, and promising results have already been obtained. Figures of merit for repeat analyses of a mixture of test compounds that are not dissimilar from those observed in HPLC analysis have been reported. Robson *et al.*[52] showed that, with both unpressurized and pressurized systems, highly repeatable separations can be obtained; for a series of injections of a test mixture, relative standard deviations (RSDs) were less than 1% for retention time and, typically, 1–3% for peak height and corrected peak area. In addition, retention time, column efficiency and retention factor have been demonstrated to remain essentially constant for at least 200 repeat injections on the same column.[38,52]

Still lacking, however, are convincing demonstrations of trace analysis for, say, impurities at the 0.1% level. Longer light-path flow cells are becoming available, and it may be that current experiments with low (or even zero!) electrolyte concentrations, and hence reduced ohmic heating, will permit larger column diameters to be utilized.

# References

1. H.H. Strain, *J. Am. Chem. Soc.*, 1939, **61**, 1292.
2. C. Haugaard and T.D. Kroner, *J. Am. Chem. Soc.*, 1948, **70**, 2135.
3. H.H. Strain and J.C. Sullivan, *Anal. Chem.*, 1951, **23**, 816.
4. D.L. Mould and R.L.M. Synge, *Biochem. J.*, 1954, **58**, 571.
5. J. Kowalczyk, *J. Chromatogr.*, 1964, **14**, 411.
6. R.M. Hybarger, C.W. Tobias, and T. Vermeulen, *I. and E.C. Proc. Design Develop.*, 1963, **2**, 65.
7. G. Bundschuh, *J. Chromatogr.*, 1971, **56**, 241.

8. J. Salak and P. Roch, *J. Chromatogr.*, 1972, **71**, 459.
9. P.H. O'Farrell, *Science*, 1984, **277**, 1586.
10. V. Pretorius, B.J. Hopkins, and J.D. Schieke, *J. Chromatogr.*, 1974, **99**, 23.
11. J.W. Jorgenson and K.D. Lukacs, *J. Chromatogr.*, 1981, **218**, 209.
12. T. Tsuda, *Anal. Chem.*, 1987, **59**, 521.
13. J.H. Knox and I.H. Grant, *Chromatographia*, 1987, **24**, 135.
14. J.H. Knox and I.H. Grant, *Chromatographia*, 1991, **32**, 317.
15. N.W. Smith and M.B. Evans, *Chromatographia*, 1994, **38**, 649.
16. N.W. Smith and M.B. Evans, *Chromatographia*, 1995, **41**, 197.
17. R.J. Boughtflower, T. Underwood, and C.J. Paterson, *Chromatographia*, 1995, **40**, 329.
18. M.M. Robson, M.G. Cikalo, P. Myers, M.R. Euerby, and K.D. Bartle, *J. Microcol. Sep.*, 1997, **9**, 357.
19. M.G. Cikalo, K.D. Bartle, M.M. Robson, P. Myers, and M.R. Euerby, *Analyst*, 1998, **123**, 87R.
20. L.A. Colon, K.J. Reynolds, R. Alicea-Maldonado, and A. Fermier, *Electrophoresis*, 1997, **18**, 2162.
21. E.R. Verheij, U.R. Tjaden, W.M.A. Niessen, and J. van der Greef, *J. Chromatogr.*, 1991, **554**, 339.
22. D.J. Shaw, 'Electrophoresis', Academic Press, London, 1969.
23. J.H. Knox, *Chromatographia*, 1988, **26**, 329.
24. C.L. Rice and R. Whitehead, *J. Phys. Chem.*, 1965, **69**, 4017.
25. J.A. Taylor and E.S. Yeung, *Anal. Chem.*, 1993, **65**, 2928.
26. T. Tsuda, M. Ikedo, G. Jones, R. Dadoo, and R.N. Zare, *J. Chromatogr.*, 1993, **632**, 201.
27. J.C. Giddings, 'Unified Separation Science', Wiley, New York, 1991.
28. M.M. Dittman and G.P. Rozing, *J. Microcol. Sep.*, 1997, **9**, 399.
29. T.S. Stevens and H.J. Cortes, *Anal. Chem.*, 1983, **55**, 1365.
30. Q.-H. Wan, *J. Phys. Chem. B.*, 1997, **101**, 8449.
31. M.G. Cikalo, K.D. Bartle, and P. Myers, *J. Chromatogr. A*, 1999, **836**, 25.
32. A.S. Rathore and Cs. Horvath, *J. Chromatogr.*, 1996, **743**, 231.
33. Cs. Horvath and W.R. Melander, in 'Chromatography, Part A', ed. E. Heftmann, (Journal of Chromatography Library, Vol 22A), Elsevier, Amsterdam, 1983, p. A27.
34. J.H. Knox, *J. Chromatogr.*, 1994, **680**, 3.
35. M.J.E. Golay, in 'Gas Chromatography', ed. D.H. Desty, Butterworth, London, 1958.
36. G.J.M. Bruin, P.P.H. Tock, J.C. Kraak, and H. Poppe, *J. Chromatogr.*, 1990, **517**, 557.
37. Cs. Horvath and H. Lin, *J. Chromatogr.*, 1978, **149**, 43.
38. M.M. Dittman, K. Wienand, F. Bek, and G.P. Rozing, *LC-GC*, 1995, **13**, 800.
39. M.G. Cikalo, K.D. Bartle, and P. Myers, *J. Chromatogr. A*, 1999, **836**, 35.
40. S. Lüdtke, T. Adam, and K.K. Unger, *J. Chromatogr. A.*, 1997, **786**, 229.
41. S.C.P. Roulin, K.D. Bartle, M.R. Euerby, and C.M. Johnson, unpublished measurements.
42. R.J. Hunter, 'Zeta Potential in Colloid Science', Academic Press, London, 1981, p. 61.
43. P.B. Wright, A.S. Lister, and J.G. Dorsey, *Anal. Chem.*, 1997, **69**, 3251.
44. M.R. Euerby, D. Gilligan, C.M. Johnson, S.C.P. Roulin, P. Myers, and K.D. Bartle, *J. Microcol. Sep.*, 1997, **9**, 373.
45. K. Sealey, K.D. Bartle, and P. Myers, unpublished measurements.
46. N.M. Djordjevic, P.W.J. Fowler, F. Houdiere, and G. Lerch, *J. Liq. Chrom. Relat. Technol.*, 1998, **21**, 2219.
47. B. Chankvetadze, 'Capillary Electrophoresis in Chiral Analysis', John Wiley, Chichester, 1997, Ch. 9, p. 353.
48. G. Choudhary and Cs. Horvath, *J. Chromatogr. A*, 1997, **781**, 161.

49. E. Grushka, *Anal. Chem.*, 1970, **42**, 1142.
50. P. Huang, J.-T. Wu, and D.M. Lubman, *Anal. Chem.*, 1998, **70**, 3003.
51. M.L. Lee, F.J. Yang, and K.D. Bartle, 'Open Tubular Column Gas Chromatography', Wiley-Interscience, New York, 1984.
52. M.M. Robson, S.C.P. Roulin, S.M. Shariff, M.W. Raynor, K.D. Bartle, A.A. Clifford, P. Myers, M.R. Euerby, and C.M. Johnson, *Chromatographia*, 1996, **43**, 313.

CHAPTER 2

# *The Capillary Electrochromatograph*

NORMAN W. SMITH

## 1 Introduction

Electrochromatography can be performed using standard commercial capillary zone electrophoresis (CZE) equipment or equipment that has been self-assembled. Although several workers[1-3] have reported performing CEC using equipment without pressurization it is noteworthy that in all instances the full operating potential of CEC was not being exploited *i.e.* the analyses were being carried out at $\leqslant 20$ kV often at below ambient temperature (15–20 °C). This is almost certainly to avoid bubble formation and it was Knox[4,5] who pointed out that this can arise in CEC through self-heating and the fact that the packing materials themselves act as excellent nuclei for bubble formation. The out-gassing appears to occur at the outlet frit and is generally recognize as resulting from differences in EOF between the packed bed region of the capillary and the open tube region.

Since high buffer concentrations can give rise to a significant improvement in efficiency, it is desirable to work with electrolytes at concentrations up to 0.02 M. However, Knox reported problems with bubble formation at electrolyte concentrations $> 6 \times 10^{-3}$ M. The way to overcome this problem, and allow the use of high-concentration buffers, is to pressurize the whole column as suggested in the same paper and this forms the basis of the modified equipment described in this chapter.

## 2 Modification of Standard CZE Equipment

Until 1996, there were no commercial instruments available that allowed CEC to be studied on a pressurized system. Since 1996, three instrument manufacturers have produced pressurized equipment. These manufacturers are:

- Hewlett Packard (HP $^{3D}$CEC)

- Beckman (MDQ) and
- Thermoseparations (Ultra).

The properties of these instruments are described later (Section 5).

Prior to these instruments appearing, CEC carried out under pressure and at elevated temperatures could only be achieved by modification of existing CZE equipment. Smith[6,7] modified two pieces of standard CZE equipment, an ABI 270A and a Unicam Lauerlabs Prince instrument. (All modifications were carried out by the Bio-engineering Department of Glaxo-Wellcome Research & Development Limited, Stevenage, UK, according to the author's designs and specifications. Before these instruments could be modified the corresponding instrument manufacturers were required to disengage autosamplers and 'trick' the instrument software so that manual injections could be performed, and the equipment ran under pressure.) In the case of the ABI instrument, the autosampler and autosampler arm had to be removed and in the case of the Unicam instrument the 'cathode' end of the system had to be completely revamped. Figure 2.1 is a schematic layout of the pressurized system whilst Table 2.1 is a list of the corresponding components used for the modification.

Once the packed capillary had been inserted into the corresponding cell, Kel-f nuts were placed over the capillary ends along with 1/16″ Vespel ferrules. The cells were then locked in place and the capillary ends threaded through the Kel-f end fittings (item 4, Figure 2.2) into the buffer reservoirs and then tightened in place.

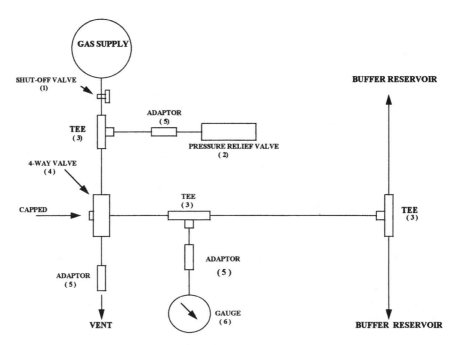

**Figure 2.1**  *Pressurization system for EC*

**Table 2.1** *List of components*

| Drawing number (see Figure 2.1) | Part description | Code number | Supplier | Number required |
|---|---|---|---|---|
| 1 | Peek injection moulded shut-off valve | P.733 | Upchurch | 1 off |
| 2 | Pressure relief valve | B-4CPA2-350 | North London Valve Company | 1 off |
| 3 | 3-Way block | 1055 | Omnifit | 3 off |
| 4 | 4-Port valve | 1121 | Omnifit | 1 off |
| 5 | Adaptor | | Bioengineering | 1 off |
| 6 | Pressure gauge | 111.10.40 | Wika Instruments | 1 off |

**Figure 2.2** *Components of the pressurization vessel: (1) Tefzel end cap (top); (2) Kel-F bottom fitting; (3) glass column (50 mm × 17 mm i.d.); (4) Kel-F top fitting; (5) O-ring; (6) Tefzel end cap (bottom)*

Figure 2.1 shows that the pressurization system was connected to an air cylinder which was usually set to approximately 200 psi. A pressure relief valve (2), set to open at 500 psi was inserted into the system to prevent accidental over-pressurization. The pressure was applied and released by way of a 4-way valve (4). A pressure gauge (6) was incorporated into the system via an adapter

(5) and by connection to a tee (3). The pressure line was split *via* another tee and connected to the two buffer reservoirs which consisted of 50 mm × 17 mm i.d. glass Omnifit chromatography columns capable of withstanding pressures up to 1000 psi.

A more detailed picture of the components of these pressurized vessels is shown in Figure 2.2. The components are shown approximately full size. Although the pressure relief valve was designed to open at 500 psi, an operating pressure of 100–200 psi was found to be sufficient to prevent bubble formation. Initially, the end fittings of the pressurized glass columns, which at one end are designed to take a capillary and electrode, and at the other the pressurized gas, were constructed from poly(tetrafluoroethylene) (PTFE). However, this material proved unsuitable mainly because of creep when working under pressure. When this was replaced by Kel-F, these fittings were found to be entirely trouble-free.

The modified ABI and Unicam instruments respectively are shown in Figures 2.3 and 2.4.

These modified instruments were to prove extremely reliable and had the added bonus of being able to accept capillaries of different lengths, unlike their commercial counterparts where the capillary dimensions are far more rigid (a drawback of the cassette system!). The major disadvantage of the modified CZE instruments is the fact that they are manual instruments and therefore only single sample analysis is possible. The procedure for sample injection is therefore relatively lengthy.

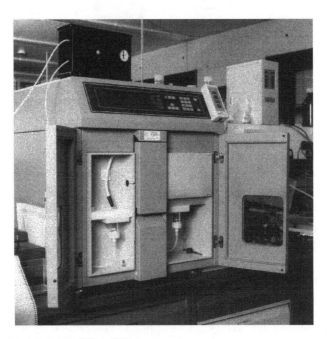

**Figure 2.3**   *Modified ABI 270A CZE instrument*

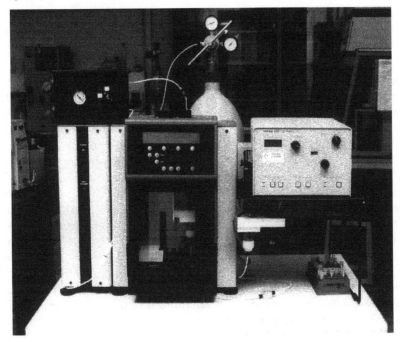

**Figure 2.4** *Modified Unicam Prince CZE instrument*

# 3 Sample Injection Using Modified CZE Equipment

With packed capillaries, electrokinetic injection is the generally preferred
method of sample introduction, although some workers have used the hydrodynamic technique. Because of the small diameter packings used and the length
of the packed beds, hydrodynamic injections are usually made for long time
periods because the pressure available for such injections is insufficient considering the high column back pressure. An electrokinetic injection procedure
has been developed involving the following steps:

    (i) switch off the voltage;
   (ii) release the pressure;
  (iii) dismantle the pressure vessel on the injection side (anode) and remove
       the buffer vial;
  (iv) reconnect the pressure vessel with the injection block containing the
       sample solution;
   (v) apply pressure (typically 15 bar) – optional;
  (vi) turn on the voltage for a set period (typically 10 kV for 0.5 min);
 (vii) switch off the voltage;
(viii) release the pressure;
  (ix) dismantle the pressure vessel and remove the sample vial;
   (x) reconnect the pressure vessel with the buffer vial inserted;

    (xi) apply the pressure;
    (xii) switch on the voltage.

In the event of a capillary developing air bubbles (indicated by an unsteady current and noisy baseline) it is necessary to purge the capillary using a high-pressure pump *e.g.* an HPLC column packing pump (which is a pressure intensifier pump) or a HPLC pump. This is achieved by removing the capillary plus cell from the instrument and connecting them to the pumping system using a 1/16″ union and a 1/16″ Vespel ferrule. When the column packing pump is used, the operating pressure is usually set to about 6000 psi, and water is pumped through the system. With the HPLC pump the flow of water is usually set such that the back pressure remained below the upper pressure limit of the pump. If the pressure exceeded the upper limit, the pump would cut out and the purging process would stop. Flow rates in the region of 0.06–0.1 ml min$^{-1}$ usually gave a back pressure in the region of 4000 psi. Removal of bubbles using these procedures is usually complete within 15–30 min. The pressure is then carefully released and the packed capillaries placed back in the instrument. A capillary rarely develops air bubbles when running under pressurized conditions with these modified instruments.

# 4 Modification of CZE Equipment with Autosampler Capability

It is possible to modify existing CZE equipment to work under pressure and perform automated sample analysis. Several workers[8–12] have carried out this modification by attaching a gradient HPLC system to a CZE instrument and pressurizing the packed capillary via some form of splitting device. A typical layout is shown in Figure 2.5.

Although this modification has been demonstrated to work well, and transforms standard CZE equipment into an automated system (assuming the HPLC equipment has autosampler capability), it suffers from quite long dwell times because the capillary connecting the HPLC instrument to the CZE equipment is prohibitively long. If a modular HPLC system is used, however, it is possible to configure the pump and autosampler closer to the CZE instrument, making shorter connections to the 1/16″ union (3) and 1/16″ union tee (5) and reducing the dwell time significantly. Samples are introduced into the system by applying a flow of ~100–200 μl min$^{-1}$ using the HPLC pump. The vast majority of the sample passes through the by-pass capillary (6) because of the resistance of the packed column. However, a small amount of sample will enter the packed bed by a combination of EOF and pressurized flow. Outgassing is prevented by introducing another restrictor at the outlet end of the packed capillary (11). By careful assembly of the inlet capillary to the 1/16″ union tee (5) efficiencies equal to those generated on commercial instruments were easily achieved.

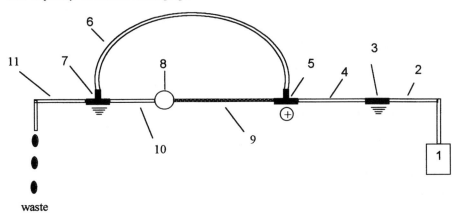

**Figure 2.5** *Modified CZE instrument using a gradient HPLC pump Components: (1) Gradient HPLC system; (2), (4) Connecting capillary 50 μm i.d.; (3) 1/16" stainless steel union; (5), (7) 1/16" stainless steel union tee; (6) By-pass capillary 50–100 μm i.d.; (8) Detector cell; (9) Packed capillary; (10) Unpacked length of capillary; (11) Waste capillary*

## 5 Commercial CEC Instruments

In the early days of CEC, work was conducted on modified CZE equipment. CEC requires voltages up to 30 kV (1 kV per cm of column), producing currents between ~0.5 and 30 μA, and these conditions could be adequately achieved by instruments such as the ABI 270A (Foster City, CA, USA) and the Unicam ATI Prince (Cambridge, UK). Recently instrument manufacturers have realised the potential of CEC and have designed instruments to perform automated CZE and CEC. All of these instruments apply pressure to both inlet and outlet ends of the capillary. The three main contenders in this field, Beckman Instruments, Hewlett Packard and ThermoQuest, have now produced automated CEC equipment.

Beckman Instruments launched their CE/CEC instrument – the MDQ – in 1997. The MDQ has (2 × 36) 2 ml buffer vials allowing 36 different pairs of mobile phase reservoirs to be used with their standard tray, in any sequence. The sample tray can take a variety of sample configurations including 2 × 96 well plates or (2 × 48) 2 ml vials. Samples can be stored in an optional temperature controlled environment from 5 to 60 °C. The MDQ uses a cartridge system with a standard capillary packed length of 20 cm. However, if longer cartridge tubing is fitted (through which the coolant circulates), extended CEC capillary lengths can be used. The length of unpacked capillary from the detection window to the outlet buffer vial is 10 cm. Pressures of up to 100 psi can be applied to both ends of the capillary to suppress bubble formation. Electrokinetic injections can be performed by applying up to 10 kV for a specified time. Hydrodynamic injections can also be made by applying a pressure of 100 psi to the inlet vial. The MDQ uses a recirculating liquid coolant system, which has a working range of 15 to 60 °C. Detection is either by diode array (190–600 nm) or ultraviolet–visible (UV–Vis) using standard filters of 200, 214, 254, and 280 nm, or 190–600 nm with custom filter options.

The Hewlett Packard $^{3D}$CE instrument uses a self-aligning cartridge system that is air cooled, thus simplifying the overall design. The cartridge uses, as standard, capillaries of packed length 24.5 cm with an additional 8.5 cm from the detector to the outlet vial. Pressurization of both ends of the capillary up to 12 bar is possible. A fan-assisted oven controls the temperature of capillaries from 10 °C below ambient up to 60 °C. Samples and mobile phase vials are stored in a 48-position carousel which can be temperature controlled from 10 to 40 °C using an external water bath. As with the Beckman MDQ, samples can be injected electrokinetically. However, with this instrument voltages up to 30 kV can be applied. Hydrodynamic sample injection is also possible by applying up to 12 bar to the inlet vial only, but because of the high column pressure resistance, injections need to be made over long time periods. Buffer depletion during long sample sequences can be overcome by use of an automated off-line buffer replenishment unit. This allows the use of valuable positions in the sample carousel not required for multiple buffer vials.

Detection is by a diode-array specifically designed for on-capillary detection. Sensitivity can be further enhanced by use of an extended light-path cell, but although this has been successfully used in the CZE mode, practical difficulties have restricted its used in CEC.

The contribution to the market from ThermoQuest has been the Ultra CE. This instrument allows for smaller capillaries to be used, packed length ∼20 cm with an additional 6.5 cm from the detection window to the outlet buffer. This overall length of ∼26 cm is short compared to the other two instruments and has the potential advantage of working at higher field strengths. Pre-aligned optics enable the cassette to be installed very quickly, and a unique bar-code label automatically logs the amount of usage, times, and dates, and maintains a log of average voltages and currents. Injection can be made electrokinetically at voltages up to 30 kV or hydrodynamically by applying 100 psi to the inlet end. Outgassing is prevented by the application of 100 psi to both ends of the capillary. The Ultra temperature is Peltier controlled from 10 °C below ambient to 60 °C and can operate in the constant-temperature or temperature-gradient mode. The autosampler has a capacity for 110 vials and a sample tray which can also be Peltier controlled to between 0 and 60 °C (optional). Detection is provided by a scanning UV–Vis detector with a dual ball-lens design to enhance sensitivity. The scanning detector allows spectral analysis as well as contour and three-dimensional (3-D) plots.

All three commercial instruments have built in sophisticated data handling and system control providing total system manipulation and data reprocessing. However, none has the ability to run at voltages higher than 30 kV.

## 6  Future Instrument Designs and Requirements

No instrument currently available has been specifically designed to run CEC. All of the instruments described in section 5 are CZE instruments modified to run under pressure, thus allowing CEC to be performed. What is needed is an instrument designed to accommodate CZE, CEC and micro-HPLC. Shared components would make the cost considerably less than the sum of the three individual machines. Also, all three techniques could be used without having to switch instruments and, in the case of micro-HPLC and CEC, without having to switch capillaries! It would be desirable if such an instrument was capable of running at voltages greater than 30 kV, but how high the limit should be is uncertain at the moment because of arcing and other problems under such conditions.

Although the cassette design has many desirable attributes, it can be very restrictive and the whole concept of how best to house the capillary, whilst giving flexibility in the length of the packed bed for example, needs addressing. Pressurization is probably the biggest issue currently confronting instrument manufacturers. None of the manufacturers of CEC instruments to date has a foolproof system that allows pressurization of the packed capillary, which is so essential for the successful implementation of this technique. This could of course be solved by implementing the design outlined in Section 4, whereby the HPLC pump is used to pressurize the system. Another consideration for the manufacturers is to make such instruments mass-spectrometer compatible. This should ideally be achieved without the standard type of interface, which uses connecting tubing to take the sample into the mass spectrometer, because such designs always lead to a dramatic decrease in efficiency. Lane *et al.*[13] showed that it was possible to design a CEC–mass-spectrometer interface without an open tube section thus losing no efficiency due to connecting tubing. This also eliminates problems associated with bubble formation, which tends to occur at the interface between the packed and open regions of the capillary. Their design also meant that because short capillaries could be used, so could high field strengths leading to very fast analyses.

Although CEC can progress with existing equipment, its acceptance as a serious analytical technique will increase substantially with the advent of purpose-built machines, preferably of the 'triskelion type', which can perform three functions, i.e. CZE, CEC and micro-HPLC.

# References

1. C. Yan, R.J. Dadoo, H. Zhao, and R.D. Zare, *Anal. Chem.*, 1995, **67**, 2026.
2. M.T. Dulay, C. Yan, D.J. Rakestraw, and R.N. Zare, *J. Chromatogr.*, 1996, **725**, 361.
3. M.M. Robson, S. Roulin, S.M. Shariff, M.W. Raynor, K.D. Bartle, A.A. Clifford, P. Myers, M.R. Euerby, and C.M. Johnson, *Chromatographia*, 1996, **43**, 313.
4. J.H. Knox and I.H. Grant, *Chromatographia*, 1991, **32**, 317.
5. J.H. Knox, *Chromatographia*, 1988, **26**, 329.
6. N.W. Smith and M.B. Evans, *Chromatographia*, 1994, **38**, 649.
7. N.W. Smith and M.B. Evans, *Chromatographia*, 1995, **41**, 197.
8. M.R. Taylor and P. Teale, *J. Chromatogr. A*, 1997, **768**, 89.
9. M.R. Taylor, P. Teale, and S.A. Westwood, *Anal. Chem.*, 1997, **69**, 2554.
10. C. Yan, R. Dadoo, R.N. Zare, D.J. Rakestraw, and D.S. Anex, *Anal. Chem.*, 1996, **68**, 2726.

11. R.J. Boughtflower, personal communication, 1998.
12. B. Behnke and E. Bayer, *J. Chromatogr.*, 1994, **680**, 93.
13. S.J. Lane, R.J. Boughtflower, C. Paterson, and M. Morris, *Rapid Commun. Mass Spectrom.*, 1996, **10**, 733.

CHAPTER 3

# Supports and Stationary Phases for Capillary Electrochromatography

PETER MYERS

## 1 Introduction

In this chapter we present a review of the main support materials and bonded phases that are used in capillary electrochromatography.

In the first papers published on CEC[1,2] the materials used were commercially available supports that had been developed for HPLC, commonly being 5 $\mu$m silica based. These materials had pore sizes of 8–10 nm, with pore volumes of 0.4–0.7 ml g$^{-1}$ and surface areas of 150–300 m$^2$ g$^{-1}$. ODS was the most common bonded phase. However, the true particle size distribution and the bonding silane were never reported. Later discussions will show how these play a very important role in CEC where the particle and bonded phase are used both to generate the flow and partition the analytes.

Later papers have used smaller diameter silica particles[3,4] and recently non-porous silica particles from 0.5 $\mu$m to 1.5 $\mu$m have been used.[5–9] In all of these papers a general reduction in particle size has been observed. However, great care has to be taken when comparing the effect of particle size, as the solid particles in the particle size range 0.5–1.5 $\mu$m are generally manufactured by the Stoeber process[10] and so are monodispersed with a far higher density than porous silica particles. As discussed later, this can have pronounced effects on the packing of these particles into the narrow tubes used in CEC.

The original porous particles were usually manufactured by the sol–gel process. A description of this manufacturing method and its many variants is outside the scope of this chapter, but full details can be found in the classic book by Ralph Iler.[11] The point that must be made is that all the commercially available silicas and other supports are manufactured by different methods and so will have different surface and pore characteristics. From our work[12] these differences are far more significant in CEC than HPLC.

When comparing methods using different silicas, the manufacturing method should, if possible, be noted. The older type silicas such as Spherisorb and Hypersil are manufactured from a sol–gel process, but of these two silicas one is manufactured by a reaction-limited coagulation and the other by diffusion-limited coagulation, so two very different silicas are produced. Both of these older silicas also contain large amounts of sodium, as their starting material is a sodium-stabilized sol. In the case of Spherisorb sodium is present at a level of 1500 ppm.

The newer cleaner silicas such as Symmetry are manufactured by the hydrolysis and polycondensation of tetraethoxysilane (TEOS). Even more recently, hybrid silicas that have methyl groups embedded into the silica matrix have been introduced. All these silicas have very different surface properties and physical parameters that in CEC effect and control both the EOF and the separation of analytes.

To the author's knowledge no one has yet researched the full optimization of the pore size, pore volume, surface area and silica type for CEC. The only research that has been reported is that by Li and Remcho[13] who reported results on pore size *versus* efficiency. This research must be carried out if CEC is to be developed and its full mechanism understood.

## 2   Particle Size Distributions

An important difference between porous particles from different manufacturers is their particle size distribution. The three more common ways of displaying the particle size is by number, area and volume. The number mean diameter is given by equation (3.1),

$$X_{nl} = \frac{\Sigma dl}{\Sigma dN} = \frac{\Sigma dn}{\Sigma dN} \qquad (3.1)$$

the area mean diameter by equation (3.2)

$$X_{na} = \sqrt{\frac{\Sigma dl}{\Sigma dN}} = \sqrt{\frac{\Sigma x^2 dn}{\Sigma dN}} \qquad (3.2)$$

and the volume mean diameter by equation (3.3)

$$X_{nv} = \sqrt{\frac{\Sigma dV}{\Sigma dN}} = \sqrt{\frac{\Sigma x^3 dn}{\Sigma dn}} \qquad (3.3)$$

It is common in HPLC to discuss the particle size distribution as an area or a volume distribution, both of which give a larger figure than the number distribution. They also hide the very fine particles as in the number distribution each particle has the same weighting, whereas in the area distribution this is a squared function such that a 10 $\mu$m particle has 100 times the weighting of a 1 $\mu$m and in a volume distribution the 10 $\mu$m particle has a 1000 weighting compared with the 1 $\mu$m particle. This is shown in Figure 3.1.

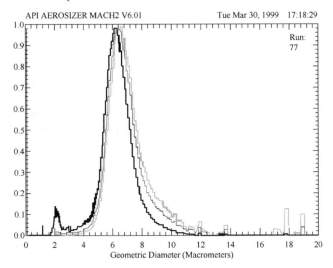

Figure 3.1 *Number, area and volume distributions for an HPLC-grade silica*

In HPLC the particle size determines both the column back pressure and its efficiency. But what is the effect of the particle size in CEC? Results show that the efficiency does scale in the same way as in HPLC but that small fine particles can fill in the interstitial voids between the larger particles to such an extent that they produce double layer overlap and hence stop or reduce the plug flow profile. In the materials that were specially developed for CEC by Phase Separations great care was taken to remove all the fine material.

# 3 The Silica Surface

Although the characterization of the surface of the silica is outside the scope of this chapter and book, in very general terms the silica surface can be considered to comprise silanols, SiOH, and siloxane bridges, SiOSi. Within this classification there are several types of silanols, single, geminal and bridged (also called vicinal). These are illustrated in Figure 3.2, while Figure 3.3 shows a typical

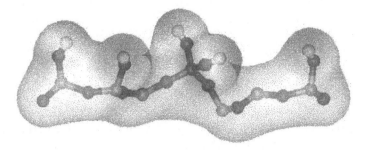

Figure 3.2 *Model showing vicinal, geminal and single silanols*

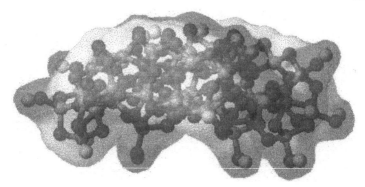

**Figure 3.3**   *Model of a typical silica surface*

silica surface. In most research papers the discussion about silica assumes that it is pure; no work in chromatography has taken into account the fact that silica contains many impurities, such as sodium, iron, etc, and that these metal ions are complexed into the silica matrix. No one has shown how these metals affect the ion-exchange properties of the silica or how different manufacturing methods affect the ratio of these metals in the silica.

Silanols are acidic and many papers have been written on the pH of a fully hydroxylated surface. As a general rule it is taken that the pH is between 3 and 4. At high pH the silanols dissociate leaving a negative surface. This is the surface that forms the double layer that in turn leads to the EOF. There is much debate about the degree of acidity of each type of silanol, how metal impurities incorporated into the matrix contribute to this acidity and what is the total concentration of silanols on the silica surface. Published values range from 3 silanol groups $nm^{-2}$ to 12 groups $nm^{-2}$. Again the actual value is very silica dependent; it depends on the manufacturing method and on the conditioning or pretreatment of the silica. Much more research is needed to understand and optimize the silica surface for use in CEC.

## 4   Bonding Technology

The majority of silica bonded phases as used today in both HPLC and CEC are based on the silicon–carbon bond and this is derived through the reaction of silica with silanes. A typical reaction is to react a chlorosilane with a silanol on the surface of silica. Figure 3.4 shows the simplest reaction scheme but it is drawn with a solvent surface and shows the electrostatic potential around the silanol. However, there are a vast number of reactive silanes containing one, two and three reactive groups, which can be chloro- or alkoxysilanes. No manufacturer appears to specify the type of silane that is used. The percentage carbon loading of the silane is commonly given as a w/w figure of the carbon but this does not give the information that is critical in CEC, namely the surface coverage of available silanols.

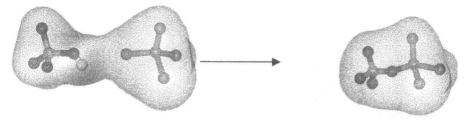

**Figure 3.4** *Simple reaction of a monofunctional silane with an isolated silanol*

When a monofunctional silane is reacted with silica only about 2.5 silanols $nm^{-2}$ can be captured, which means that well over half of the silanols remain unbonded. This is generally assumed to be due to steric hindrance, although this author disagrees. In an attempt to capture all the remaining silanols an endcapping technique is used where a small silane is bonded onto the silica. The bonded phase materials that have been developed for CEC are in general classed as non-endcapped materials. An attempt has been made[14] to review the electroosmotic mobility of different stationary phases against different material types. Only part of a very complex picture could be presented as a full characterization would require the full details relating to the silica and silane used by the manufacturers.

If a difunctional silane is used it might not be possible for both functional groups to react with surface silanols. In this case a new silanol will be produced on the ligand after the functional group is hydrolysed. The original silanol ligand density would then be barely different from the silanol density on the silica or on a monofunctional silane. With a trifunctional silane that only bonds to a one surface silanol, there are two functional groups that can be hydrolysed into silanols, so the silanol activity can be increased in this case.

Another problem with di- and trifunctional silanes is that they can polymerize through a self-reaction between the silanes with one silane bonding onto the reactive hydrolysed silanol of another silane, so leading to the so-called polymeric phases.

## 5 Effect of pH on the Silanols

If the pH of the silanols is taken to be in the order of pH 4–5 and these are assumed to be the driving force of the EOF then as the pH of the mobile phase is reduced then so will the EOF, as the surface of the silica is no longer charged. In an attempt to increase the EOF at low pH, sulfonic acid phases were introduced. In the first reported work[15] a propylsulfonic acid was bonded onto the surface of a very narrow particle size 5 $\mu$m Spherisorb silica. Results from a number of workers[16,17] have shown that the EOF is maintained as the pH is dropped. But the most remarkable results were obtained with this phase and reported by Smith and Evans[15] in their work on the separation of tricyclic antidepressants such as amitriptyline and nortriptyline. They reported extremely high efficiencies in the order of millions of plates. Their results have now been confirmed by others[18] and also the problems in obtaining these very high efficiencies in a reproducible manner. The focusing effect has also been reported by other workers[19] with other phases for both anionic and neutral analytes.

# 6   Analyte Focusing

## Propylsulfonic Acid Phases

In our own work we have tried to investigate and understand the propylsulfonic acid phase and this has been done using molecular modelling.

Using Hyperchem[20] molecular modelling software a solvated model of a section of the propylsulfonic acid (SCX) chromatographic surface was created. The system consisted of a slab of silica gel with several propylsulfonic acid chains attached which were randomly solvated with mobile phase molecules (Figure 3.5). Obviously, this model contained hundreds of atoms and tens of molecules and so optimization using Monte Carlo (MC) techniques would be an impossibly long process, and even using molecular modelling (MM) would be highly laborious. By setting up the model so that the silica surface is fixed in space and ignored by the calculations, the geometry and orientations of the propylsulfonic acid chain bonded to the silica along with the solvent molecules can be optimized more rapidly. On this model, using MM optimization, 1000

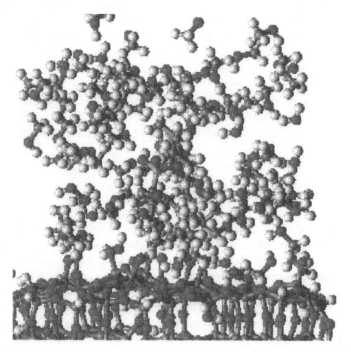

**Figure 3.5**   *The complex water:methanol solvated system of a single propylsulfonic acid bonded to a silica surface*

**Figure 3.6** *Optimized model with an energy of* $-487.6 \, kcal \, mol^{-1}$

iterations took approximately 24 hours to complete on a 300 MHz Pentium processor.

Mobile phases of water, methanol and a 50:50 mixture of water:methanol were simulated by positioning the molecules around the silica and propylsulfonic acid using an Excel macro that randomly orientated a determined number of a designated molecule within a selected 3-D boundary. By varying the water content in the mobile phase, different minimum energy configurations of the sulfonic acid have been found. At high aqueous concentrations it is found that the propylsulfonic acid lies parallel to the silica surface. In performing these calculations the solvent molecules were first optimized by MC and then the whole system by MM to give the change in orientations given in Figure 3.6.

When the propylsulfonic acid was then modelled in a high organic mobile phase, *i.e.* the organic component was >60%, the propylsulfonic acid was found to stand perpendicular to the silica surface and to be solvated by methanol molecules. Using the Excel macro the optimization produced a much lower energy change, again with little movement of the stationary phase as shown in Figure 3.7. In this configuration the organic species cluster around the propyl chain and appear to act as a 'collar' in keeping the ligand perpendicular to the silica surface.

## Other Sulfonic Acid Phases

Wei *et al.*[21] have attached a polymer layer containing sulfonic acid groups onto silica. Their reported efficiencies were in the range 40–90 $\times$ $10^3$ plates m$^{-1}$ and the peak shape was not good. Further work is therefore under way to investigate this polymer layer and to try to understand if the sulfonic acid can move under different organic strengths and to determine whether it is this effect that leads to the focusing.

**Figure 3.7**  *Optimized model with an energy of $-491.5 \, kcal \, mol^{-1}$*

There have been reports of mixed mode $C_8$ sulfonic acid phases. Again comparisons are very difficult to make as they have been on different silicas.

# 7   Anion Exchange Materials

To complement the SCX materials strong anion exchange (SAX) materials have also been manufactured and bonded onto silica. A number of groups have reported different separations[22,23] including the separation of myoglobin and the enantiomeric separation of N-derivatized amino acids. Our own work[24] has shown that the EOF generated by this phase is very dependent on the SAX bonding and on the packing of the bonded phase into the column. Under some conditions an EOF was generated in the same direction as that for a pure silica column, although the expected flow would be in the reverse direction.

# 8   Continuous Bed Columns

A number of papers have been published describing continuous bed column.[25–28] These have been manufactured by a variety of different routes ranging from the sintering of packed spherical particles to the utilization of TEOS to form a continuous bed inside the column. Researchers appear to have been forced down the latter route through the unavailability of small spherical silica particles. Silica manufacturers do not seem to be very willing to supply loose packing materials. The problem with TEOS continuous bed columns is that every column is a new individual batch of material. For years silica manufacturers have been trying to manufacture reproducible silica. Today they are a lot better than in the early 1980s but part of this control is by manufacturing such large batches of silica that over 1000 HPLC columns can be manufactured from one batch. Using the continuous bed techniques we could be back to vast column to column reproducibility problems.

# 9 Conclusions

This review highlights the urgent need for research into supports and bonded phases that are specifically designed for CEC. To date the packings used have by default been those used in HPLC. However, before these new supports and phases can be developed, a greater understanding of the mechanisms involved in CEC is needed. This relates to the packing and design of columns, the flow profiles obtained and the connection of CEC columns to instruments.

The new supports and phases should be capable of being packed into capillary tubes to provide channels that will not induce double layer overlap, so as to maintain a plug profile. The new materials may well incorporate permanent changes to enhance the EOF across a wide range of pH conditions and they will include new and novel phases to enhance both selectivity and efficiency. Quite a challenge.

# References

1. W.J. Jorgenson and K.D. Lukacs, *J. Chromatogr.*, 1981, 218.
2. J.H. Knox and I.H. Grant, *Chromatographia*, 1987, **24**, 135.
3. N.W. Smith and M.B. Evans, *Chromatographia*, 1994, **38**, 649.
4. M.M. Robson, S. Roulin, S.M. Shariff, M.W. Raynor, K.D. Bartle, A.A. Clifford, P. Myers, M.R. Euerby, and C.M. Johnson, *Chromatographia*, 1996, **43**, 313.
5. B. Behnke, E. Grom, and E. Bayer, *J. Chromatogr. A*, 1995, **716**, 207.
6. R.M. Seifar, W.T. Kok, J.C. Kraak, and H. Poppe, *Chromatographia*, 1997, **46**, 131.
7. H. Engelhardt, S. Lamotte, and F.T. Hafner, *Am. Lab.*, 1998, **30**, 40.
8. R. Dadoo, R.N. Zare, C. Yan, and D.S. Anex, *Anal. Chem.*, 1998, **70**, 4787.
9. C.G. Bailey and C. Yan, *Anal. Chem.*, 1998, **70**, 3275.
10. S. Lildtke, T. Adams, and K.K. Unger, *J. Chromatogr. A*, 1997, **786**, 229.
11. Ralph K. Iler, 'The Chemistry of Silica', John Wiley & Sons.
12. M.G. Cikalo, K.D. Bartle, and P. Myers, *Anal. Chem.*, 1999, **71**, 1820.
13. D.M. Li and V.T. Remcho, *J. Microcol. Sep.*, 1997, **9**, 389.
14. T.M. Zimina, R.M. Smith, and P. Myers, *J. Chromatogr.*, 1997, **758**, 191.
15. N.W. Smith and M.B. Evans, *Chromatographia*, 1995, **41**, 197.
16. M.Q. Zhang and Z. El Rassi, *Electrophoresis*, 1998, **19**, 2068.
17. N. Smith and M.B. Evans, *J. Chromatogr. A*, 1999, **832**, 41.
18. M.R. Euerby, D. Gilligan, C.M. Johnson, S.C.P. Roulin, P. Myers, and K.D. Bartle, *J. Microcol. Sep.*, 1997, **9**, 373.
19. F. Moffatt, P.A. Cooper, and K.M. Jessop, *Anal. Chem.*, 1999, **71**, 1119.
20. HyperChem, 1115 NW 4th Street, Gainesville, FL 32601, USA.
21. W. Wei, G.A. Luo, and C. Yan, *Am. Lab.*, 1998, **30**, 20C.
22. M. Lammerhofer and W. Lindner, *J. Chromatogr. A*, 1998, **829**, 115.
23. D.M. Li, H.H. Knobel, and V.T. Remcho, *J. Chromatogr. B*, 1997, **695**, 169.
24. K. Sealy, PhD Thesis, University of Leeds, 1999.
25. A. Palm and M.V. Novotny, *Anal. Chem.*, 1997, **69**, 4499.
26. N. Chen, C. Ericson, and S. Hjerten, *Anal. Chem.*, 1996, **68**, 3468.
27. C. Fujimoto, J. Kino, and H. Sawada, *J.Chromatogr. A*, 1995, **716**, 107.
28. J.L. Liao, E.C. Peters, M. Petro, F. Svec, and J.M.J. Frechet, *Anal. Chem.*, 1997, **69**, 3646.

CHAPTER 4

# Electroosmosis in Complex Media: Bulk Transport in CEC

## VINCENT T. REMCHO AND PATRICK T. VALLANO

## Preface

Capillary electrochromatography (CEC) is a powerful separation technique that essentially combines the advantages of capillary HPLC and capillary electrophoresis (CE). As in capillary HPLC, solutes partition between the mobile and stationary phases. However, instead of a pressure gradient to pump the mobile phase, application of an electric field to the capillary induces electroosmotic flow (EOF), which is responsible for bulk transport. The favorable flow dynamics of EOF translate into high chromatographic efficiency in CEC separations. Partitioning as a separation mechanism combined with EOF as a means of bulk transport imparts high selectivity as well as efficiency to CEC. Thus resolving power in CEC, in theory, should be higher than in capillary HPLC or CE.

Presently, CEC is most commonly performed in the packed column format using spherical, silica-based reversed-phase particles designed for HPLC. Relative to open tubes, electroosmosis in these packed beds is much more complex. The purpose of this chapter is to discuss EOF in packed column CEC, with specific reference to phenomena that may cause deviations from the assumed 'plug' flow profile and degrade efficiency. In addition, alternatives to conventional packed CEC, including monolithic beds and large diameter, wide-pore media are discussed.

## 1  Introduction

CEC is a separation technique that can be viewed as a variant of HPLC, in which bulk transport of the mobile phase is achieved by electroosmosis instead of pressure. In a typical CEC experiment, an electric field is applied axially to a packed fused silica capillary column. Electroosmotically driven mobile phase transports solutes through the column, where partitioning or another separative process occurs between phases. Charged species will have an additional, electrophoretic velocity component that may contribute to or impede the separation. In this sense, CEC can be considered a hybrid that combines the features of capillary zone electrophoresis (CZE) and capillary HPLC. As a high peak capacity technique, CEC is particularly suited for analyzing complex mixtures, such as biological or environmental samples.

The driving force in the development of CEC has been the ability to achieve high resolving power and peak capacity. The superior performance of CEC with respect to these figures of merit is attributable primarily to the favorable dispersive characteristics of EOF, which are manifest in lower plate heights (higher efficiencies) relative to pressure driven flow. This chapter will explore EOF in packed capillary electrochromatography.

# 2   Electroosmotic Flow

## The Electrical Double Layer: Origin of Electroosmosis

The electrical double layer exists at solid–liquid interfaces as a consequence of a charge separation between the solid surface and the adjacent solution. When in contact with aqueous solutions, most solids acquire a surface electrical charge by mechanisms such as ion adsorption from the bulk solution or direct ionization of functional groups at the solid surface. For example silica, including fused silica capillary tubing and derivatized particulate materials employed in CEC, acquires a net negative charge due to dissociation of acidic silanol (SiOH) moieties at its surface. When the siliceous material is placed in contact with an aqueous solution, a charge imbalance exists owing to the negatively charged surface and an excess of positively charged counterions present in the bulk solution that arise from the dissociation of protons. Electrostatic forces result in the attraction of the excess counterions to the surface. The charged surface and counterions constitute a 'double layer' that consists of negative and positive charges. The formation of such a double layer at a fused silica surface is depicted schematically in Figure 1.1 on page 3. The blanket of counterions near the solid surface consists of two regions. The layer nearer to the surface, the Stern layer, is composed of a monolayer of immobile ions. Farther from the surface, a dynamic equilibrium exists between the forces of electrostatic attraction and thermal motion that results in the formation of a layer of mobile counterions that can freely exchange with ions in the bulk solution, termed the diffuse or Gouy region of the double layer. The formation of the Gouy layer results in a charge imbalance between it and the bulk solution. In order to gain an understanding of the variables that control EOF, it is instructive to begin with a more detailed examination of the electrical double layer.

The formation of the electrical double layer gives rise to a potential that varies as a function of distance from the particle surface. This relationship is indicated in Figure 4.1, in which potential ($\psi$) *versus* distance is plotted. With the aid of this figure, several features of the double layer can be noted. The potential at the

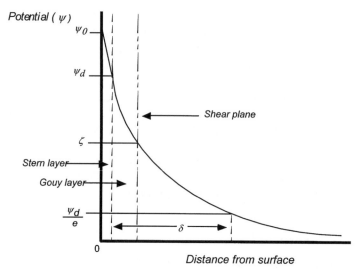

**Figure 4.1**   *Potential as a function of distance from a charged surface. The potential at the plane of shear is the zeta potential, $\zeta$. The distance over which the potential at the boundary of the Stern and Gouy layers, $\psi_d$, decays by a factor of $1/e$ is known as the double layer thickness, $\delta$*
(Adapted from ref. 2)

surface ($\psi_0$) is dependent on the surface charge density (which in turn depends in part on the extent of silanol dissociation on a silica surface). Potential decreases with distance in a linear fashion across the Stern layer. At the interface between the Stern and Gouy layers, the potential is indicated as $\psi_d$. Across the diffuse region and into the bulk solution, the decay in potential is roughly exponential ($\psi \rightarrow 0$ as $x \rightarrow \infty$). The distance over which $\psi_d$ decays to $1/e$ of its original value is denoted by $\delta$, and is termed the 'double layer thickness' (alternatively expressed as $\kappa^{-1}$). Double layer thickness is given by the following equation, which can be derived by means of the Poisson–Boltzmann distribution and the Debye–Hückel approximation:[1]

$$\delta = \left( \frac{\epsilon_0 \epsilon_r RT}{2F^2 c z^2} \right)^{1/2} \tag{4.1}$$

where $\epsilon_0$ = permittivity of vacuum, $\epsilon_r$ = dielectric constant of the solution, $R$ = gas constant, $T$ = temperature (Kelvin), $F$ = Faraday's constant, $c$ = electrolyte concentration and $z$ = electrolyte valence. As indicated in equation (4.1), increasing the concentration (ionic strength) of the electrolyte in the solution will compress the double layer, a relationship of importance in CEC.

In the diffuse region of the double layer, a plane of shear exists that separates those species not free to exchange with the bulk solution (hydrated counterions) from free species. The potential at this shear plane is termed the zeta potential ($\zeta$). The zeta potential is dependent upon double layer thickness as:[2]

$$\zeta = \frac{\delta\sigma}{\epsilon_0\epsilon_r} \tag{4.2}$$

in which $\sigma$ = charge density at the shear surface (other variables defined previously). Substitution of equation (4.1) into equation (4.2) yields

$$\zeta = \frac{\left[\left(\dfrac{\epsilon_0\epsilon_r RT}{2F^2 cz^2}\right)^{1/2}\right]\sigma}{\epsilon_0\epsilon_r} \tag{4.3}$$

Values of $\delta$ are on the order of nm and typical values for a monovalent electrolyte are $\delta = 10$ nm for $c = 10^{-3}$ M and $\delta = 1$ nm for $c = 10^{-1}$ M. If all other values are substituted into the equation, including reasonable approximations of $\epsilon_r$ and $T$, it is easily determined that typical zeta potentials fall in the 25–100 mV range for silica surfaces and aqueous solutions.

## Electroosmotic Flow

When an electric field is applied along the length of a fused silica capillary filled with an aqueous solution, a force is exerted on the excess counterions in the Gouy layer, and as a result they begin to slip along the shear plane and migrate toward the oppositely charged electrode. As these hydrated ions migrate, they drag along with them (as a result of appreciable frictional forces) the bulk solution, resulting in electroosmotic flow. Because no charge imbalance exists in the bulk solution, the electroosmotic flow profile is flat. When the field is applied, equilibrium is quickly established in which viscous forces balance electrostatic forces. Equating these forces, the following expression, known as the Smoluchowski equation, can be derived.[1]

$$\mu_{eo} = \left(\frac{\varepsilon_0\varepsilon_r\zeta}{\eta}\right)E \tag{4.4}$$

where $\mu_{eo}$ is the electroosmotic flow velocity, $\eta$ = solution viscosity and $E$ is the applied field strength. This equation is valid only in the absence of double layer overlap, a phenomenon that occurs in small diameter flow channels (treated in later sections). Note the linear dependence of $\mu_{eo}$ on applied field strength, $E$ (*i.e.* a plot of $\mu_{eo}$ *versus* $E$ yields a straight line for a broad range of field strengths). Depending on the conditions employed, slightly curved plots are sometimes observed, which may indicate thermal effects. These effects can degrade efficiency, as discussed later in this chapter. It is of particular importance in equation (4.4) that there is no dependence of velocity on flow channel diameter.

## Operational Variables

### *Field Strength*

Equation (4.4) shows the linear relationship between EOF velocity and applied field strength. In practice, varying field strength is a useful means by which to manipulate flow velocity in CEC separations. It is often desirable to operate at high field strengths in order to provide rapid analysis times; however, practical limitations are imposed by Joule heating effects and by power supplies, which usually cannot exceed 30 kV.

### *Ionic Strength*

As predicted by equations (4.1–4.4), EOF velocity decreases with increasing ionic strength of the mobile phase (ionic strength $= \frac{1}{2}\sum_i c_i z^2$). At high ionic strengths, compression of the double layer results in a decrease in the magnitude of the EOF. This affords a mechanism for gross control of bulk flow velocity, fine control being exerted by alteration of the electric field strength.

### *pH*

The dissociation equilibrium of the surface silanol groups ($pK_a$ 2–4) is pH dependent. The surface charge density, and hence the zeta potential, is a function of the extent of ionization. As expected, EOF velocity increases on going from acidic to slightly alkaline pH owing to increased silanol dissociation. However, little or no change is observed at higher pH values, at which most of the surface silanol groups are dissociated.[3] Given the relatively modest range over which changes in pH affect EOF velocity, and the profound effect that pH may have on the solution behavior of certain analytes, it seems most prudent to settle on a choice of pH that is dictated by the sample.

### *Organic solvents*

The type and percentage composition of organic solvent in the mobile phase can affect the magnitude of the EOF. The relationship is a complex one: organic modifiers may affect EOF through changes in bulk properties of the mobile phase (*e.g.* dielectric strength and viscosity) and by altering the zeta potential.

In bare fused silica capillaries, it has been found that EOF velocity decreases upon addition of acetonitrile to an aqueous buffer.[4] In packed column CEC, contradictory results have been obtained in which EOF velocity was found to increase[5,6] and decrease[7,8] with increasing volume per cent of organic solvent in the mobile phase. Although differences in packing materials and buffer compositions employed in these studies may partially explain the inconsistency, these results indicate the complexity of the relationship.

Interestingly, it has been shown that generation of EOF is possible in CEC with neat non-aqueous solvents.[9–12] Comparatively robust EOF velocities were obtained with pure acetonitrile, which has been shown to yield an estimated zeta potential of over 200 mV in a bare silica capillary.[11] At present, the mechanism by which double layer formation occurs in non-aqueous solvents is unclear.

# 3 Electroosmosis in Packed Beds

## Peak Dispersion

Longitudinal broadening of solute bands (peaks) is the inevitable result of dispersive processes that occur on passage through the chromatographic column. Because chromatographic separations are based on the differential migration of solute zones, peak dispersion acts in opposition to the separative transport process and ultimately limits resolution. In chromatography, the extent of peak dispersion can be expressed by the height equivalent to a theoretical plate (HETP, $H$), or simply plate height. Plate height, expressed is length units, is defined as the peak variance per unit length of column.

$$H = \frac{\sigma_l^2}{L} \qquad (4.5)$$

In this equation, $\sigma_l^2$ = peak variance (in units of length) and $L$ = column length.

The principal causes of chromatographic peak dispersion are molecular diffusion, sorption–desorption kinetics and flow anisotropy.[13] Typically, flow anisotropy is the dominant dispersive phenomenon near the minimum achievable plate height ($H_{min}$) in packed column chromatography. Point variations in flow velocity occur within the column owing to several phenomena. Within a single flow channel, in an interstitial space for example, radial non-uniformity in flow velocity leads to peak dispersion. In pressure-driven systems (*e.g.* in HPLC), flow velocity at any radial point in the channel is a function of distance from the center. Peak dispersion arises from this effect in the following way. In a given time interval, solute molecules near the center of the channel (where flow velocity is at its maximum, $\mu_{max}$) will outdistance the zone as a whole, which travels at the average velocity $\frac{1}{2}\mu_{max}$. Likewise, molecules near the wall (where $\mu = 0$) will fall behind with respect to the average. As a consequence, the solute band as a whole is broadened. Radial molecular diffusion, however, allows molecules to sample the various velocity regimes, which provides an averaging effect, counteracting dispersion. Unfortunately, the relatively slow diffusion rates of molecules in liquids limit this diffusional averaging.

Additional sources of flow anisotropy exist within packed columns. In contrast to an open tube, the pathways available for fluid flow within a packed chromatographic column consist of an intricate network of interconnected channels, highly variable in geometry and connectivity. For this reason, flow through chromatographic media is difficult to characterize and approximations must be made (*e.g.* the capillaric model). In traversing a fixed length increment of column, for example, some molecules may travel *via* a more direct pathway, while others may encounter obstructions and thus travel an indirect route. The molecules composing the zone as a whole will then have varying velocity components along the column axis that are distributed about some mean value. The resultant peak, on passing through the detection cell, will appear axially broadened.

Variability in the geometry of flow channels within a single column exists as a consequence of the random nature of the packing structure. For example, assuming flow occurs only within the interstitial spaces in a chromatographic column packed with uniformly sized spheres, the geometry of the flow channels will be dependent upon the porosity ($\varepsilon$) of the packed bed. Porosity is defined as the fraction of free space in the column with another fraction $(1 - \varepsilon)$ being occupied by the solid packing material. The compactness of the packing particles will determine the porosity. For instance, the minimum porosity of close packed perfect sphere is achieved with a rhombohedral arrangement, for which $\varepsilon = 0.26$. A less compact arrangement is the simple cubic array ($\varepsilon = 0.48$).[14] An average interstitial porosity of 0.4, between these extremes, is generally assumed for chromatographic packings.[13] In addition to particle diameter, the diameter of the interstitial flow channels is related to the porosity of the packed bed.[24] During column packing the arrangement of packing particles is random and the orientation of particles with respect to each other (*i.e.* the packing density) can vary considerably.[13] Thus, within the column, point-to-point differences in values of local porosity (*i.e.* the porosity within any small volume element) will exist. As a consequence, there will be non-uniformity in the effective flow channel diameter (*i.e.* some channels will be more permeable to flow than others). Using again the example of pressure-driven flow, because flow velocity is proportional to $1/d_{ch}^2$, small variations in channel diameter ($d_{ch}$) can result in relatively large differences in flow velocity. In general, zone broadening phenomena that arise from the structural properties of the packed bed are termed eddy diffusion effects.

The Knox equation is generally employed in packed column liquid chromatography to assess the dependence of $H$ on various dispersive processes.[15] Expressed in reduced parameters, the Knox equation is:

$$h = Av^{1/3} + \frac{B}{v} + Cv \qquad (4.6)$$

Where $h$ = reduced plate height ($h = H/d_p$), $v$ = reduced velocity ($v = \mu d_p/D_m$), $d_p$ = particle diameter, $D_m$ = solute diffusion coefficient in the mobile phase, $\mu$ = mobile phase linear velocity, $A$ = constant representing the coupled effects of eddy diffusion and resistance to mass transfer in the mobile phase, $B$ = constant representing axial molecular diffusion and $C$ = constant representing resistance to mass transfer in the stationary phase. The $A$ term of the Knox equation is a relative measure of how well packed a column is. A poorly packed column will, in general, have local porosity values that fluctuate over a wider range than a well packed column. The result will be a larger $A$ term and a correspondingly higher reduced plate height. The $A$ term cannot, in principle, be eliminated in packed column chromatography (pathways of varying complexity will invariably be present); however, it can be minimized by reducing variations in flow velocity that arise from a non-uniform packing structure. In HPLC, this can be accomplished in part by the use of packing techniques that yield more homogeneous beds.

## Advantages of EOF in Packed Column Chromatography

Electroosmosis is inherently advantageous as a means of driving the mobile phase in chromatography. The favorable flow dynamics of EOF reduce peak dispersion and enable higher efficiencies to be obtained relative to pressure-driven flow. Theoretically, there are three features of EOF that can be exploited in packed column CEC to yield increased efficiency and ultimately higher resolving power: (1) EOF velocity is independent of channel diameter over a broad range encompassing current conventional channel diameters, (2) EOF velocity is independent of radial position in a single channel assuming reasonable selection of operational parameters and (3) EOF requires no pressure drop.

### *Eddy Diffusion*

In packed column chromatography, the most important advantage conferred by electroosmosis is the independence of flow velocity on channel diameter. Previously, the contribution of packing non-uniformity to peak dispersion through the dependence of (pressure-driven) velocity on channel diameter was discussed. EOF velocity, however, is essentially independent of channel diameter;* thus variations in flow velocity between regions of the column differing in packing structure will be comparatively small in an electrically driven system, as depicted schematically in Figure 4.2. In the absence of double layer overlap, a uniform velocity distribution would be expected throughout the column, notwithstanding differences in packing density. As a result, overall plate height in the electrically driven system will be lower, owing to a smaller $A$ term.

This intrinsic advantage of EOF as it relates to packed column chromatography was demonstrated as long ago as 1974 by Pretorius *et al.*[16] More recently, Knox and Grant[17] evaluated identical columns with pressure and electrically driven flow and obtained lower plate heights in the CEC mode. Values of the $A$ term for both modes of flow were obtained from $h$ *versus* $v$ plots.[18] Not surprisingly, the estimated eddy diffusion coefficient was lower in the electrically driven mode ($A_{\mu LC} = 0.8$ and $A_{CEC} = 0.3$), a direct consequence of the more uniform EOF velocity profile.

### *Transchannel Velocity Uniformity*

In the absence of double layer overlap, EOF velocity is independent of radial position (except very close to the wall) and is given by equation (4.4). Pressure-driven flow, in contrast, has a velocity profile that varies in a parabolic fashion

---

* In the limit of no double layer overlap or excessively large flow channels.

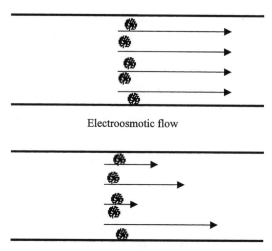

Electroosmotic flow

Pressure driven flow

**Figure 4.2** *Representation of flow velocity profiles through channels of varying diameter. (Arrows represent mean velocity vectors.) Unlike pressure-driven flow, EOF velocity is independent of channel diameter. Thus, a uniform velocity profile is obtained even when flow channels vary in size*

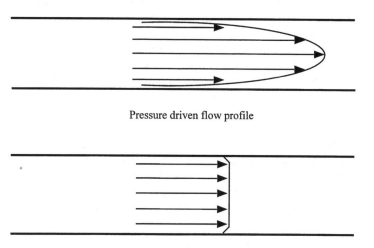

Pressure driven flow profile

Electroosmotic flow profile

**Figure 4.3** *Comparison of transchannel velocity profiles. Pressure-driven flow velocity is a function of radial position, varying from a maximum at the channel center to zero at the wall. EOF velocity is independent of radial position in the channel except in the region very close to the wall. Thus, EOF velocity is essentially uniform across the channel*

with radial position. Pressure- and electro-driven flow profiles are depicted schematically in Figure 4.3.

The parabolic variation in flow velocity results in peak broadening due to slow mass transfer, as solute diffusion is insufficiently rapid for cross-channel averaging. In contrast, this transchannel effect will be substantially less in electrically driven systems because the flow velocity across any given channel is essentially uniform.

## *Pressure Drop*

In chromatography, the relationship between plate height and particle diameter is well known.[13,19] However, the use of small particles ($d_p < 3$ $\mu$m) to maximize efficiency in HPLC is limited by the magnitude of the pressure drop required to drive mobile phase through the substantially smaller flow channels. (The pressure drop, $\Delta P$, needed to achieve a given flow velocity is proportional to $d_p^{-2}$.) EOF imposes no such limitation; thus CEC is amenable to the use of small diameter packing media.

## Geometric Constraints on EOF in Packed Beds

Other variables constant (*i.e.* $\varepsilon$, $\eta$, $\zeta$, $E$ and column diameter), flow velocity will be lower in magnitude in a packed capillary than in an open tube. This is due in part to non-alignment of most flow channels with the electric field, which is applied along the column axis. Consider a capillary tube packed with solid (impermeable) spherical particles. In such a tube, flow will occur only within the interparticulate spaces. As noted previously, the individual flow channels comprise a highly complex, directionally non-uniform network, of which only a fraction will be aligned axially with the electric field. Most flow channels will be off-axis with respect to the field and as such will experience a lower effective field strength. As pointed out by Knox,[17] the effective field strength in any flow channel will be $E \cos \theta$ where $E$ is the applied field strength and $\theta$ is the angle between the tube and channel axes. In addition to experiencing a lower effective field strength, the fluid flowing in an off-axis channel must travel a greater distance (increased by a factor of $1/\cos \theta$) for a given displacement along the column axis. The combination of these factors, shown pictorially in Figure 4.4, results in the reduction of the velocity component along the tube axis by a factor of $\cos^2 \theta$.

The silica-based particulate materials commonly employed in CEC are porous. Although too narrow, in general, to support electroosmosis, the pore spaces are occupied by fluid and can be sampled by analyte. Thus, the flow velocity averaged over the cross-sectional area permeable to fluid will be lower for a porous than for a nonporous, packing medium. Specifically, the flow velocity measured in a porous packed bed (*e.g.* measured by the migration time of a neutral, unretained solute) will be lower than the 'true' velocity by a factor dependent on intraparticle porosity.[17,20] That is, the greater the porosity of the particles themselves, the lower the flow velocity.

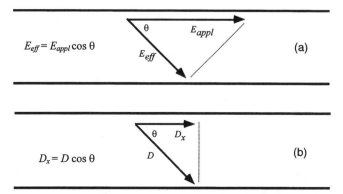

**Figure 4.4** *Non-alignment of a flow channel with column axis. (a) The potential gradient (electric field) is applied along the column axis (x). If the magnitude of the field along x is denoted by $E_{appl}$, then the effective field along any channel deviating from the column axis by an angle θ ($E_{eff}$) will be $E_{appl}$ cos θ. Thus, $E_{eff} < E_{app}$ when the channel is not aligned with the field axis (i.e. when θ ≠ 0°). (b) For any distance (D) traveled in a channel oriented at an angle θ with respect to the column axis, the displacement along the column axis, $D_x$, will be D cos θ. The cumulative result of factors (a) and (b) is a reduced velocity component along the column axis by a factor of $cos^2$ θ*

It has been estimated that the combined factors of geometric non-alignment with the field axis and intraparticle porosity result in a 40–60% reduction in EOF velocity in porous packed beds relative to open capillaries.[17] However, the predominance of packed column CEC over its open tubular counterpart indicates that the velocity reduction does not impose a serious limitation on the technique.

## Other Factors Affecting Efficiency in Packed Column CEC

### Double Layer Overlap

The preceding discussion of EOF has assumed that the flow channel is large relative to double layer thickness. It is only in this limiting case that the Smoluchowski equation (equation 4.4) is valid. When the double layer thickness approaches the radius of the flow channel, however, overlap of the double layers may occur, which results in diminished EOF velocity and a degradation of the flow profile into a Poiseuille-like distribution. In the extreme, EOF velocity approaches zero.

A theoretical paper by Rice and Whitehead[21] on electroosmotic flow in capillary tubes studied the flow profile distortion that arises due to double layer overlap. In it, they derived the following relationship between EOF velocity and radial position in the tube:

$$\mu(r) = \frac{\varepsilon_0 \varepsilon_r \zeta}{\eta} E \left[ 1 - \frac{I_0(\kappa r)}{I_0(\kappa a)} \right] \qquad (4.7)$$

where $\mu$ = flow velocity, $r$ = distance from the tube center, $I_0$ is a zero order modified Bessel function of the first kind, $\kappa$ is the reciprocal of the double layer thickness (refer to equation 4.1) and $a$ = capillary radius. The product $\kappa a$, termed the electrokinetic radius, is the ratio of the capillary radius to the double layer thickness. It can be shown that when $\kappa a$ is large, the term in brackets approaches unity and EOF velocity is essentially independent of radial position [in this case, equation (4.7) reduces to the Smoluchowski equation]. Plots of the transchannel velocity profile for several values of $d_{ch}/\delta$ ($d_{ch}/\delta$ = $\kappa a$, $d_{ch}$ = interstitial flow channel diameter) are shown in Figure 4.5. It can be seen that significant deviations from the flat profile are observed for low values of $\kappa a$ owing to overlap of the double layers extending from the channel walls. It is important to note from this plot that when $\kappa a$ = 20, EOF velocity is uniform over 90% of the channel. Relative to open tubular (OT) techniques such as OTCEC and CZE, double layer overlap is more probable in packed column CEC owing to the fact that the flow channels are much narrower. Interstitial channel diameters are a factor of approximately 0.20–0.30 of particle diameter, depending upon the estimation method.[22–24] Under current typical operating conditions, double layer overlap in CEC is not a major problem. As particle diameter (or *effective* particle diameter) is decreased, however, greater attention must be paid to this phenomenon.

Eliminating or at least minimizing double layer overlap in CEC is important because the decrease in EOF velocity and the flow profile distortion that occur can substantially degrade efficiency. When overlap occurs, point differences in velocity are introduced, increasing overall plate height. As discussed previously, non-uniform flow channels arise from local differences in packing density. If the interstitial flow channels (diameter = $d_{ch}$) in a more densely packed region of the column are sufficiently small such that the ratio $d_{ch}/\delta$ falls below about 10, a uniform velocity profile can no longer be assumed and efficiency will be decreased.

In a paper published early in the development of CEC, Stevens and Cortes[22] reported significant reductions in EOF velocity with 10 $\mu$m diameter ($d_p$ = 10 $\mu$m) particles (relative to the velocity observed with 100 $\mu$m particles)

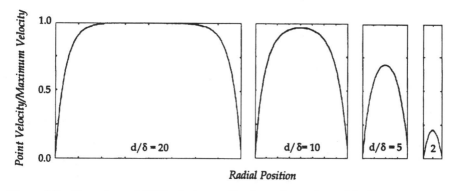

**Figure 4.5**   *Transchannel EOF velocity profiles for various values of* $d_{ch}/\delta$

which they attributed to double layer overlap. Knox and Grant, however, demonstrated theoretically[23] and experimentally[17] that double layer overlap effects were minimal for the types of eluents and packing materials typically employed in CEC. Their experimental studies showed that EOF velocity is independent of particle size down to $d_p = 1.5$ $\mu$m, contradicting earlier conclusions.

Table 4.1 shows estimates of the minimum concentrations of electrolyte that can be used to maintain a $d_{ch}/\delta$ value of at least 20 with various sizes of particles. When this condition is satisfied, flow velocity will be uniform over 90% of the channel.[21] Inspection of Table 4.1 shows that the minimum concentrations in each case are quite low. In the extreme case, for 1.5 $\mu$m particles, a minimum concentration of 0.00027 M can be tolerated. Particles smaller than 1.5 $\mu$m could be used with higher minimum concentrations.

## Thermal Effects

The CEC capillary is analogous to a cylindrical conductor: when the electric field is applied, current flows and resistive (Joule) heat is produced. The quantity of heat generated per unit volume of electrolyte ($Q$) within the column is given by the following expression:[25]

$$Q = E^2 \lambda \, c\varepsilon \tag{4.8}$$

where $E$ = field strength, $\lambda$ = molar conductivity of the electrolyte, $c$ = electrolyte concentration and $\varepsilon$ = total porosity. This heat is dissipated by conduction through the capillary wall and surrounding air resulting in a parabolic temperature gradient across the capillary bore, the magnitude of which is given by:[25]

$$\Delta T = \frac{Qd_c^2}{16\kappa} = \frac{E^2 \gamma c\varepsilon}{d_c^2 16\kappa} \tag{4.9}$$

in which $\Delta T$ = temperature difference between the center of the capillary and the inner wall, $d_c$ = capillary diameter and $\kappa$ = thermal conductivity of the mobile phase. This temperature difference will lead to a viscosity gradient

**Table 4.1** *Minimum allowable electrolyte concentration to achieve* $d_{ch}/\delta = 20$ *with various particle sizes*

| $dp/\mu m$ | $d_{ch}/nm$ | $\delta_{max}/nm$ | $C_{min}/mmol\ L^{-1}$ |
|---|---|---|---|
| 10 | 2500 | 125 | $6.0 \times 10^{-3}$ |
| 5 | 1250 | 62.5 | $2.4 \times 10^{-2}$ |
| 3 | 750 | 37.5 | $6.7 \times 10^{-2}$ |
| 1.5 | 375 | 18.8 | $2.7 \times 10^{-1}$ |

Assumed conditions: monovalent (1:1) electrolyte; $\varepsilon_r = 80$ (water); $T = 25\ °C$; $d_{ch} = 0.25\ d_p$.

between the capillary core and the inner wall. The net result will be a parabolic variation in flow velocity across the capillary radius (recall that EOF velocity is inversely proportional to mobile phase viscosity).

Knox[25] derived an expression for the contribution of thermal effects to plate height in CEC.

$$H_{TH} = 10^{-8} \left( \frac{\varepsilon_0 r \delta}{D_m \eta \kappa^2} \right) E^5 d c^6 \gamma^2 c^2 \qquad (4.10)$$

In this equation, $H_{TH}$ is the plate height increment due to thermal effects and $D_m$ is the diffusion coefficient of the solute in the mobile phase (other variables defined previously). From this equation, we see that $H_{TH}$ is large with wider capillaries, higher electrolyte concentrations and high field strengths. Typically, $H_{TH}$ is insignificant for column diameters less than 100 $\mu$m.

The random packing structure also has a bearing on efficiency loss arising from thermal effects. Because the heat released per unit volume is dependent upon the porosity of the packing medium (equation 4.8), heat production (and its subsequent effect on viscosity) will be greatest in regions of the column that are more densely packed. Thus, although there is no explicit dependence on EOF velocity on channel diameter (in the absence of double layer overlap), packing non-uniformity can lead to variations in local flow velocity (and a concomitant increase in plate height) through thermal as well as double layer overlap effects.

The importance of these effects in packed column CEC has been evaluated.[27] Linear velocity as a function of applied field strength was evaluated in packed and open tubular capillaries. Positive deviations from linearity were observed at high field strengths with the packed column and ascribed to heat generation within the capillary. Increased temperature will reduce mobile phase viscosity and an increase in flow velocity will result. The fact that thermal effects were observed at the low electrolyte concentrations employed is probably due to inefficient heat dissipation from the rather large i.d. capillary ($d_c = 200$ $\mu$m) that was used. The effect of electrolyte concentration on plate height was assessed in order to investigate double layer overlap. As the electrolyte concentration was increased from 0.04 to 1.0 mM, plate height decreased in excess of 10 $\mu$m, an indication that double layer overlap was affecting efficiency. It must be noted that the electrolyte concentrations employed in this study are in a relative sense quite low, and under such conditions double layer thickness would be large.

In general, double layer overlap effects in packed column CEC will be more prevalent with dilute electrolytes, small $d_p$ and/or columns that are very tightly packed (*i.e.* having a low porosity). Thermal effects will predominate at high electrolyte concentrations, high applied fields and/or with wider bore capillaries.

## Flow Velocity Inhomogeneity

Figure 4.6 illustrates a schematic of a typical packed CEC capillary. The capillary consists of two sections, a packed region held in place by inlet and

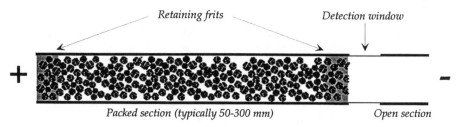

Retaining frits                              Detection window

Packed section (typically 50-300 mm)                        Open section

**Figure 4.6**   *Schematic of a typical packed capillary column in CEC. The packing material is retained by inlet and outlet frits. Detection is performed in the open section through a 'window' formed by removing a section of the polyimide cladding*

outlet retaining frits and an open section, in which detection is performed. The volumetric flow rates ($v$) through each section are given by the following:

$$v_p = \mu_p A_p \tag{4.11}$$

$$v_{ot} = \mu_{ot} A_{ot} \tag{4.12}$$

where $v$ = volumetric flow rate, $\mu$ = linear (EOF) velocity and $A$ = free cross-sectional area (in the respective sections, P = packed, OT = open tubular).

At steady state, the total fluid flux (*i.e.* the amount of fluid flowing through a cross-section of unit area per unit time) through the capillary is uniform. To fulfill this condition requires that the volumetric flow rates in the packed and open sections be equal. Thus,

$$v_p = v_{ot} \equiv \mu_p A_p = \mu_{ot} A_{ot} \tag{4.13}$$

In the majority of CEC applications, the surface of the capillary tube is untreated, resulting in a robust zeta potential. This, the tortuosity of the packed bed, and the greatly increased free cross-sectional area (*i.e.* the area not occupied by impediments, such as packing particles) in the open tubular portion results in a greater volumetric flow rate in the open section of the tube than in the packed section. Clearly for the condition in equation (4.13) to be satisfied, the flow rate in the two regions must somehow be equated by reducing the velocity in one segment and/or increasing it in the other. The mechanism by which this occurs is of interest as it applies to peak dispersion, owing to possible distortion of the EOF profile and bubble formation, a notable problem in packed column CEC.

An analogous situation arises when fluid passes through the outlet frit. Depending upon the method used for frit construction, the surface properties of the frit region, specifically the zeta potential, may vary considerably from that of the bulk packing (and that in the open section). Thus, electroosmotic flow velocity will vary in these regions.

It has been suggested by various workers that pressure gradients form in the vicinity of regions between velocity regimes.[6,27–29] For example, when the volumetric flow rate arising electroosmotically is greater in the open section than in the packed section, a negative pressure gradient is formed near the interface of the two sections and distortion of the 'plug' flow profile occurs. Velocity inequalities between the packed and open sections of the capillary have recently been the subject of detailed investigation.[28] An expression was derived for the pressure at the interface of the packed and open sections ($P_i$) in terms of the ratio of conductivity values ($\sigma$) in the packed and open sections (where $\sigma_p < \sigma_{ot}$) and the dimensionless packed length ($L_p/L_{ot}$). In the case that the volumetric flow rate in the open section exceeds that in the packed section, the overall effect of the pressure gradient is to retard the flow velocity in the open section so that volumetric flow is conserved. The generation of a pressure gradient in the open section results in a parabolic velocity variation across the capillary radius and thus a mass transfer-related contribution to plate height. According to Horváth and Rathore,[28] the plate height increment due to this distorted velocity profile can be expressed as:

$$H_{\text{open,FP}} = \frac{a^2}{24 D_m \mu_{\text{eo,open}}} \left[ \left( \frac{P_0 - P_i}{L_{\text{open}}} \right) \frac{a^2}{8\eta} \right]^2 \qquad (4.14)$$

where $H_{\text{open,FP}}$ = plate height increment due to the parabolic flow profile, $a$ = capillary radius, $D_m$ = solute diffusion coefficient in the mobile phase, $\mu_{\text{eo,open}}$ = average flow velocity in the open section, $P_0$ = pressure at the capillary inlet and outlet, $P_i$ = intersegmental pressure, $L_{\text{open}}$ = length of the open section and $\eta$ = mobile phase viscosity. Based on their calculations, $H_{\text{open,FP}}$ is significant, especially at low values of $L_p/L_{\text{tot}}$.

The formation of air bubbles is quite problematic and is recognized as a limitation of packed colum CEC.[30] Most often, bubble formation occurs at the border of the packed and unpacked sections.[6,18] A possible explanation of this phenomenon is the formation of local regions of low pressure that arise from variations in EOF velocity.[27] The heterogeneous nature of the exposed surface near the capillary outlet (packing, frit and open sections in close proximity) provides an area where non-uniformities in zeta potential, and hence EOF velocity, may exist. It is in these regions where solvent outgassing or vaporization is more probable. In practice, bubble formation is minimized by thorough degassing of the mobile phase, use of low conductivity and/or low concentrations of electrolytes (to minimize heat generation) and external pressurization of the inlet and outlet mobile phase vials.

## Perfusive Electroosmotic Flow

The packing materials typically employed in CEC are 3–5 $\mu$m porous silica beads, which are modified with an alkysilane reagent (*e.g.* ODS). Porous media are generally preferred in order to provide increased sample capacity. For the most part, the pores within these particles are quite small, ranging from about 80 to 100 Å, and thus will not, in general, support EOF owing to extreme double layer overlap in these narrow channels. As a result, in most CEC experiments flow is assumed to occur only within the interstitial region (*i.e.* the region of space between the packing particles). If particles with increased pore diameters are employed such that the extent of double layer overlap is reduced, flow through the particles augments flow around the particles. In such a case, solutes experience two distinct regions through which flow occurs (depicted in Figure 4.7): the *inter*particle (or interstitial) region and the *intra*particle region.

Intraparticulate (or perfusive) EOF is advantageous primarily from the standpoint of efficiency. In contrast to strictly interparticle flow, the effective particle diameter is smaller than the *actual* diameter when perfusion occurs. This is a direct result of the structure of most modern siliceous packing materials, which are corpuscular in nature being composed of a collection of small beads. The beads, in turn, dictate the effective particle diameter. Thus, one would expect that the plate height contributions of eddy diffusion and slow mass transfer in the mobile phase would be less when the pores are large enough to support EOF. In addition, stagnant mobile phase mass transfer effects are

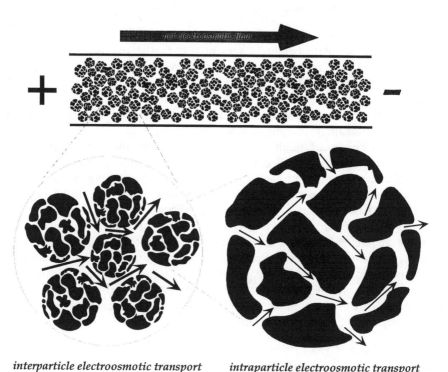

*interparticle electroosmotic transport*          **intraparticle electroosmotic transport**
                                                              *(perfusive electroosmosis)*

**Figure 4.7**    *Schematic illustrating two distinct regions in which electroosmosis may occur when capillaries packed with porous sorbents are employed in CEC. Here, bulk transport of analytes occurs in both the interparticle region (between individual particles in the packed bed) and the intraparticle region (within individual porous particles having sufficient pore diameter)*

minimized with perfusion because a greater fraction of the pores will support flow.

Perfusive electroosmotic transport in CEC was evaluated in a study by Li and Remcho[31] who employed 7 $\mu$m diameter silica particles with pore diameters up to 4000 Å. As indicated in Figure 4.8, compared to conventional particles, lower reduced plate heights were obtained with the wide-pore media. As expected, the fraction of pores large enough to support electroosmosis is greatest with the 4000 Å media. The enhanced perfusive character of the particles results in the best overall improvement in efficiency.

It is interesting to note from the data in Figure 4.8 that increases in efficiency with the wide-pore media were *not* obtained when (comparatively) dilute buffer concentrations were employed. With the 4000 Å pore diameter particles, little change in reduced plate height was observed on increasing buffer concentration from 1 to 10 mM. However, substantial changes were seen as the concentration was increased from 10 to 50 mM as reduced plate height decreased from approximately 1.85 to about 1.35. This data is consistent with perfusion. At low buffer concentrations, where double layer thickness is large relative to intraparticulate channel diameter, overlap precludes perfusion and efficiency is comparable to that of the conventional narrow-pore media. However, upon increasing buffer concentration, the double layer is compressed, yielding a more favorable (increased) $d_{ch}/\delta$ ratio and consequently increasing the perfusive character of the particles, as evidenced by the decrease in $h$.

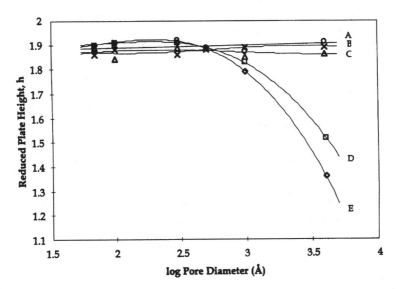

**Figure 4.8** *Plot of reduced plate height* versus *log pore diameter for a range of buffer concentrations. A = 1.0 mM; B = 5.0 mM; C = 10.0 mM; D = 30.0 mM; E = 50.0 mM. Double layer compression at high buffer concentrations allows perfusive electroosmotic transport in the wide pore media, improving efficiency* (After ref. 21)

More recently, Vallano and Remcho[32] developed a model by which the extent of intraparticle EOF could be estimated. The model allowed the estimation of an 'effective' particle diameter for various pore sizes of particles under a range of operating conditions.

The use of wide-pore media such as these with solution conditions such that perfusion occurs imparts several advantages. The enhanced mass transfer properties enable rapid separations to be obtained without a loss of efficiency. In addition, wide pores are more accessible to solutes, resulting in increased sample loadability. Lastly, packing these relatively large (5–10 $\mu$m in diameter) particles is much more straightforward than with particles below 3 $\mu$m, as is frit design, which is particularly problematic with very small particles.

# 4   Electroosmosis in Porous Monoliths

Although CEC in the packed column format provides enhanced sample loading capacity (and detection limits) relative to OTCEC, the technique has limitations. Among these are the inherent difficulties associated with packing very small particles ($d_p \leq 5$ $\mu$m) into capillary tubes, frit formation and bed stability. An alternative to packed column CEC that seeks to circumvent some of these limitations is the use of continuous polymer beds. In this approach, a monomer-containing mixture is polymerized inside the capillary to yield a macroporous, monolithic medium that can support, or itself act as, the stationery phase. Bulk flow occurs through a network of pores, the sizes of which are controlled by the polymerization conditions. Among the advantages of this approach is that the need for retaining frits is obviated. This has the added benefit that sections of the capillary can be removed and the column reused. For example, if the head of the capillary becomes fouled by a dirty sample, a few centimeters can simply be cut off, without ruining the entire column.

As with packed columns, the basic principles of chromatographic peak dispersion apply to continuous polymer beds. That is, columns should be designed such that flow anisotropy is minimized to the greatest possible extent. In monolithic beds, the effective particle diameter is dictated by the pore size. The chromatographic efficiency obtainable with a monolithic bed will depend on the ability to achieve a well-controlled, narrow pore size distribution. In addition, the pore structure must be such that mass transfer is not inhibited (*e.g.* minimizing pools of stagnant mobile phase).

Presently in CEC, continuous polymer monoliths can be divided into three general classes: (1) organic polymer, (2) inorganic (silica-based) and (3) hybrid organic–inorganic monoliths.

## Organic Monoliths

Continuous polymer beds for CEC have been prepared using various methods, the most common of which include acrylamide-[33,34] and acrylate-[35–38] based polymers. Most of these techniques require modification of the capillary surface with an anchoring reagent, to which the polymer will covalently bond.

Since 1997, rigid monoliths have been developed for CEC that required only a single step for preparation.[39–41] These acrylate-based polymers were prepared *in situ* in untreated capillaries. To yield a rigid matrix on which solutes could be retained, a hydrophobic monomer, butyl methacrylate, was combined with a high percentage of cross-linking agent. An ionizable monomer, 2-acrylamido-2-methyl-1-propanesulfonic acid (AMPS) was incorporated into the matrix to provide sufficient surface charge density for electroosmosis. A blended porogenic solvent mixture afforded tight control over pore size and breadth of pore size distribution in the monolith. Efficiencies of greater than 140 000 plates $m^{-1}$ were obtained for benzene derivatives on these monoliths.

## Inorganic Monoliths

Continuous silica-based porous monoliths have been successfully employed in HPLC.[42,43] Typically, these monoliths are prepared by a sol–gel process, which consists of polycondensation of alkoxysilanes in the presence of an organic polymer. The resultant silica skeleton contains flow-through pores with diameters on the order of a few microns. The skeleton itself is porous, with mesopores a few nanometers in diameter. The surface of the monolith is then derivatized with an organosilane reagent (*e.g.* ODS). These mesoporous properties of the monolith provide a high surface area for stationary phase bonding (and analyte partitioning). Varying the percentage of organic polymer in the mixture affords control over the porous properties of the monolith.

Silica-based continuous beds are applicable to CEC as well. Unlike the organic monoliths, incorporation of an ionizable functionality into the polymer is unnecessary in silica beds; surface silanols can support EOF. Monoliths produced *via* a sol–gel method have been employed in CEC.[44] This procedure, a variation of which has been employed in capillary gas chromatography (GC),[45] was designed to yield beds that are less susceptible to shrinking and cracking, common problems with silica monoliths.

## Hybrid Organic–Inorganic Monoliths

A procedure has been developed in which a silicate polymer was employed as an entrapment medium for polyacrylate particles.[45] In this work, the particulate material was a molecular imprint polymer (MIP) prepared with dansyl-L-phenylalanine as the imprinted species. After packing MIP particles, a solution of potassium silicate was pumped hydrodynamically into the capillary, which was subsequently heated. Polymerization of the silicate solution yielded a fine polymer network that served as a web, immobilizing the packed bed. This so-called silicate entrapment procedure has also been employed with conventional reversed-phase packing media.[45] One advantage of this type of approach with organic polymers is that the silicate matrix can support EOF; incorporation of an ionizable group into the organic polymer is unnecessary. This is especially important with MIPs, where incorporating additional monomers to provide support for EOF may interfere with the molecular recognition process.

A similar type of entrapment procedure involves the use of an organic polymer to immobilize conventional reverse-phase silica particles.[47] In addition to the intrinsic advantages of an immobilized bed (*e.g.* the elimination of frits), an organic entrapment matrix acts to minimize secondary interactions of analytes relative to a silica-based medium and affords a mechanism for fine-tuning of EOF velocity when an ionizable monomer, 2-acrylamido-2-methyl-1-propanesulfonic acid is incorporated into the mixture.

## 5 Conclusion

Electroosmosis as a means of bulk fluid transport in chromatography is potentially quite advantageous, as evidenced in the tremendous interest in CEC since the 1980s. Confirming theoretical predictions, CEC separations have been obtained in many laboratories worldwide that yield higher efficiencies than could be achieved with pressure-driven flow. However, electroosmosis lacks two of the fundamental advantages of the pressure-driven format, simplicity and reproducibility. EOF is affected by many variables in complex ways. Precise control of EOF in electrochromatographic separations is difficult for this very reason. For example, method development in an HPLC separation involves, among other factors, the choice of a suitable stationary phase and optimization of mobile phase composition. This process is inherently simpler in HPLC because flow is decoupled from these variables. The need exists, then, for a better theoretical understanding of electroosmotic flow in CEC separations, in particular in porous media. In this way, the analyst will be better equipped to exert control over electroosmosis, a necessary step in achieving in CEC the precision required by standard analytical methods.

## References

1. D.J. Shaw, 'Introduction to Colloid and Surface Chemistry', Butterworth-Heinemann, Oxford, 1992.
2. K. Mysels, 'Introduction to Colloid Chemistry', Interscience, New York, 1959.
3. M.M. Dittmann and G.P. Rozing, *J. Chromatogr. A*, 1996, **744**, 63.
4. C. Schwer and E. Kennedler, *Anal. Chem.*, 1991, **63**, 1801.
5. M.M. Dittmann and G.P. Rozing, *J. Microcolumn Sep.*, 1997, **9**, 399.
6. H. Rebsher and U. Pyell, *Chromatographia*, 1994, **38**, 737.
7. H. Yammomoto, H. Baumann, and F. Erni, *J. Chromatogr.*, 1992, **593**, 313.
8. C. Yan, D. Schaufelberger, and F. Erni, *J. Chromatogr. A*, 1994, **670**, 15.
9. J.W. Jorgenson and K.D. Lukacs, *J. Chromatogr.*, 1981, **218**, 209.
10. K.W. Whitaker and M.J. Sepaniak, *Electrophoresis*, 1994, **15**, 1341.
11. P.B. Wright, A.S. Lister, and J.G. Dorsey, *Anal. Chem.*, 1997, **69**, 3251.
12. A.S. Lister, J.G. Dorsey, and D.E Burton, *J. High Resolut. Chromatogr.*, 1997, **20**, 523.
13. J.C. Giddings, 'Dynamics of Chromatography', Marcel Dekker, New York, 1965.
14. J. Bear, 'Dynamics of Fluids in Porous Media', Dover Publications, New York, 1972.
15. J.H. Knox, *J. Chromatogr. Sci.*, 1977, **15**, 352.
16. V. Pretorius, B.J. Hopkins, and J.D. Schieke, *J. Chromatogr.*, 1974, **99**, 23.
17. J.H. Knox and I.H. Grant, *Chromatographia*, 1991, **32**, 317.

18. A.L. Crego, A. González and M.L. Marina, *Crit. Rev. Anal. Chem.*, 1996, **26**, 261.
19. J.J. van Deemter, F.J. Zuiderweg, and A. Klinkenberg, *Chem. Eng. Sci.*, 1956, **5**, 271.
20. C. Horváth and H-J. Lin, *J. Chromatogr.*, 1976, **126**, 401.
21. C.L. Rice and R. Whitehead, *J. Phys. Chem.*, 1965, **69**, 4017.
22. T.S. Stevens and H.J. Cortes, *Anal. Chem.*, 1983, **55**, 1365.
23. J.H. Knox and I.H. Grant, *Chromatographia*, 1987, **24**, 135.
24. Q-H. Wan, *Anal. Chem.*, 1997, **69**, 361.
25. J.H. Knox, *Chromatographia*, 1988, **26**, 329.
26. Q-H. Wan, *J. Chromatogr. A*, 1997, **782**, 181.
27. H. Poppe, *J. Chromatogr. A*, 1997, **778**, 3.
28. C. Horváth and A.S. Rathore, *Anal. Chem.*, 1998, **70**, 3069.
29. S. van den Bosch, Ph.D. Thesis, University of Amsterdam, 1996.
30. R.E. Majors, *LC-GC*, 1998, **16**, 12.
31. D. Li and V.T. Remcho, *J. Microcolumn Sep.*, 1997, **9**, 389.
32. P.T. Vallano and V.T. Remcho, *Anal. Chem.*, in press.
33. C. Fujimoto, Y. Fujise, and E. Matsuzawa, *Anal. Chem.*, 1996, **68**, 2753.
34. A. Palm and M.V. Novotny, *Anal. Chem.*, 1997, **69**, 4499.
35. J-L. Liao, N. Chen, C. Ericson, and S. Hjertén, *Anal. Chem.*, 1996, **68**, 3468.
36. C. Ericson, J-L. Liao, K. Nakazato, and S. Hjertén, *J. Chromatogr. A*, 1997, **767**, 33.
37. L. Schweitz, L.I. Andersson, and S. Nilsson, *Anal. Chem.*, 1997, **69**, 1179.
38. L. Schweitz, L.I. Andersson, and S. Nilsson, *J. Chromatogr. A*, 1997, **792**, 401.
39. E.C. Peters, M. Petro, F. Svec, and J.M.J. Frechet, *Anal. Chem.*, 1997, **69**, 3646.
40. E.C. Peters, M. Petro, F. Svec, and J.M.J. Frechet, *Anal. Chem.*, 1998, **70**, 2288.
41. E.C. Peters, M. Petro, F. Svec, and J.M.J. Frechet, *Anal. Chem.*, 1998, **70**, 2296.
42. H. Minakuchi, K. Nakanishi, N. Soga, N. Ishizuka, and N. Tanaka, *Anal. Chem.*, 1996, **68**, 3498.
43. N. Ishizuka, H. Minakuchi, K. Nakanishi, N. Soga, and N. Tanaka, *J. Chromatogr. A*, 1998, **797**, 133.
44. J.D. Hayes, T.J. Scott, and A. Malik, CEC 98, San Francisco, CA, USA, August 24–25, 1998.
45. J.D. Hayes, T.J. Scott, and A. Malik, *Anal. Chem.*, 1997, **69**, 3889.
46. G. Chirica and V.T. Remcho, *Electrophoresis*, 1999, **20**, 50.
47. G. Chirica and V.T. Remcho, *Anal. Chem.*, in press.

CHAPTER 5

# Capillary Electrochromatography with Open Tubular Columns (OTCEC)

MONIKA M. DITTMANN AND GERARD P. ROZING

## 1 Introduction

So far, the majority of work reported on capillary electrochromatography (CEC) has been carried out with fused silica capillaries packed with HPLC-type stationary phases. This is understandable as significant gains in separation efficiency and separation impedance are expected theoretically[1,2] if one compares a packed capillary column where the solvent is driven by hydraulic force (pressure-driven) with the electro-driven case where the solvent is driven by an electrical force. This improvement has been confirmed experimentally.[2-17] Moreover, in the electro-driven case, the solvent is dragged through the packed bed with equal force over the length of the column. Therefore, one is not penalized by increasing column backpressure, if the particle size is reduced or the column lengthened as in HPLC. With these benefits so obvious and the technological hurdles for packed column CEC in practice not too high, work in CEC has flourished since 1990.[18-48] The ultimate separation efficiency of the technique has not yet been predicted or achieved. Reports in which 1.5 $\mu$m, non-porous, reversed-phase type particles were used in CEC columns mention that 500 000 plates m$^{-1}$ have been obtained.[49]

On the other hand, open tubular liquid chromatography (OTLC) provides the best efficiency and separation impedance under conditions for liquid chromatography. Knox predicted that under optimal conditions with respect to bandspreading and provided that the internal diameter of the capillary is smaller than 5 $\mu$m, very high efficiency can be achieved.[50,51] The added benefit of an open tube is its 20–30 times higher permeability. Therefore one can afford to lengthen the capillary without being penalized with insurmountably high column pressure, as in the case for packed columns. So a column efficiency of 500 000 plates m$^{-1}$ achieved in minutes is conceivable.[51]

With these considerations in mind, one may question whether a combination of electro-driven flow with liquid chromatography in open tubes – open tubular capillary electrochromatography (OTCEC) – leads to a further improvement of the performance of OTLC. It is our intention to establish the magnitude of this improvement based on the expansion of a theoretical model we developed a few years ago.[2] This will be addressed in Section 3.

Capillary electrophoresis (CE), and in particular capillary zone electrophoresis (CZE), has become a very useful separation technique for chemical analysis.[52] The attractive aspects of the method are a simple separation mechanism, based on differences in electrophoretic mobility, high separation efficiency and putatively simple manipulation of selectivity by mere manipulation of the mobile phase (or run buffer). The capillary wall, which is fused silica as in OTLC and CEC will not, in principle, contribute to retention and selectivity of separation, but only serves as a container for the run buffer and may contribute to overall solute and solvent transport if an electro-osmotic flow (EOF) is present. However, it readily becomes clear that the capillary wall does affect separation in CE to a noticeable extent, reflected by poor reproducibility of EOF (and therefore of migration times) and in particular the shape of the eluting zone. This problem becomes particularly apparent in the CZE separation of proteins. A broad array of chemically or dynamically modified fused silica capillaries have been prepared and made available commercially to counteract this problem.

From that observation, one may argue that chromatographic-type interactions are present and play a role in CZE-type separations, affecting migration and peak shape, although these may remain unnoticed because of the low values of the capacity ratio. So in a certain sense, one may see CZE as a mode of OTCEC although with very low $k$-values. Swedberg and McManigill quantified these effects before CEC became a prominent new separation technique.[53] The magnitude of the chromatographic effects in CZE-type separations based on this work will be reviewed in Section 2.

The open tubular format of a separation column has one main disadvantage compared with packed bed capillary column, namely the very low phase ratio (ratio of the volume of stationary phase and volume of mobile phase). This will limit loadability of OTLC columns and therefore readily compromise efficiency and, in combination with the small detection volume of the technique, the limit of detection of the system. Recently new approaches have been reported in which the surface layer of the capillary is provided with some porosity, which increases the phase ratio significantly without compromising the separation efficiency. These will be discussed in Section 4.

## 2 Chromatographic Effects in CZE

In CZE, the plate height is given by:

$$H = \frac{l^2}{12L} + \frac{2D_i}{u_i} \tag{5.1}$$

in which $l$ is the length of the sample zone, $L$ is the length of the capillary column from the inlet to the detector window, $D_i$ is the diffusion coefficient and $u_i$ the velocity of the solute. The solute's velocity is given by:

$$u_i = \mu_i E + u_{eof} \tag{5.2}$$

which rearranges to

$$u_i = \frac{\mu_i V}{L_{tot}} + u_{eof} \tag{5.3}$$

in which $\mu_i$ is the electrophoretic mobility of the solute, $E$ is the field over the capillary, $V$ is the applied voltage, $u_{eof}$ is the electroosmotic flow velocity and $L_{tot}$ is the overall length of the capillary. In equation (5.1) the first term describes the plate height due to the injection of a finite length sample zone and the second term the dispersion due to static diffusion in the axial direction. Substitution of real values for the sample zone length (1 mm) and the diffusion coefficient $5 \times 10^{-5}$ cm$^2$ s$^{-1}$, at a velocity of 1 mm s$^{-1}$ yields a plate height of 2 $\mu$m which leads to 500 000 plates m$^{-1}$.

In practice, lower plate numbers are often obtained in CZE. There may be several reasons for this. Electrophoretic effects will affect bandwidth, leading to triangular peak shapes. Poor capillary thermostating, leading to radial (or axial) temperature gradients in the capillary, will cause zone broadening. Also, dispersion at sample introduction owing to an inhomogeneous field strength distribution over the capillary inlet is known to affect the bandwidth negatively.[54]

In particular with high-molecular-weight substances such as peptides and proteins or with polar, basic solutes greatly reduced plate numbers and poor peak shapes are observed in CZE. Swedberg and McManigill have attributed this to the effect of both chromatographic retention and adsorption/desorption kinetics on the plate height.[53]

Supposing that the capillary wall in CZE behaves as a retentive layer, and that the adsorption/desorption process is not infinitely fast the HETP equation has to be expanded with two terms according to Giddings:[55]

$$H_{trans} = \frac{k^2}{(1+k)^2} \frac{r_c^2 u_i}{4D_i} \tag{5.4}$$

$$H_{kin} = \frac{2k}{(1+k)^2} \frac{u_i}{k_d} \tag{5.5}$$

Equation (5.4) gives the plate height contribution of the resistance to trans-channel mass transfer when the solvent is driven by EOF. Equation (5.5) describes the plate height contribution of the adsorption/desorption kinetics. In equations (5.4) and (5.5), $k$ is the chromatographic capacity ratio, $r_c$ is the capillary radius and $k_d$ is the first order rate coefficient for the adsorption/

**HETP (µm)**

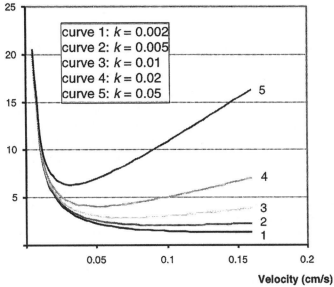

**Figure 5.1** *HETP* versus *linear velocity calculated from equation (5.6). Diffusion coefficient is $5 \times 10^{-5}$, rate constant is $10 \ s^{-1}$ (values typical for proteins)*

desorption process. In combination they lead to the following expression for the plate height:

$$H = \frac{l^2}{12L} + \frac{2D}{u} + \frac{uk}{(1+k)^2}\left[\frac{r_c^2 k}{4D_i} + \frac{2}{k_d}\right] \tag{5.6}*$$

With this expression, and substitution of meaningful values for diffusion and rate coefficients, the graphs shown in Figure 5.1 were obtained.

Figure 5.1 clearly reveals the effect of chromatographic retention on the HETP in CZE. At a velocity of 1 mm s$^{-1}$ a capacity ratio as small as 0.05 increases the HETP almost by a factor of 10 and therefore reduces the plate number by a factor of 10. If the rate constant of the adsorption/desorption process increases, the disastrous effect on HETP is mitigated, but still a factor 3 remains if the rate constant increases by a factor of 10. Lower diffusion coefficients, *e.g.* for macromolecules will also aggravate the effect.

McManigill and Swedberg have also tried to determine the relative contributions of the resistance to mass transfer and the rate constant terms to the overall HETP value obtained for a number of model protein substances. Their finding was, not unexpectedly, that for solutes with highest retention in CZE, the mass

*This equation is misprinted in the work by McManigill and Swedberg.[54] The above expression has been verified as correct with the authors.

transfer term dominates and for solutes with slow adsorption/desorption kinetics the kinetic term contributes.

# 3  Theory of Dispersion and Achievable Efficiency in OTLC and OTCEC

In the following treatment, the capillary will have a retentive layer and act as an open tubular capillary column. Two modes of operation will be conceived, *i.e.* pressure-driven and electro-driven mobile phase. In contrast to the previous section, where retention was an undesirable phenomenon, in this case retention is desired. Migration of the solutes owing to electrophoretic mobility is supposed to be absent when neutral solutes are used. Electrophoretic mobility of the solutes will complicate the description of zone dispersion. So this description applies to neutral solutes, which are separated by partitioning.

The plate height contributions for open tubular liquid chromatography are given by:

$$H = H_{diff} + H_{film} + H_{trans} + H_{stat} + H_{kin} \qquad (5.7)$$

These terms have already been described in the previous section except for $H_{film}$ and $H_{stat}$. These are the plate height increments caused by film resistance at the boundary of the retentive layer in the capillary and the contribution of resistance to mass transport by diffusion through the layer of stationary phase. The contribution by $H_{film}$ can be neglected. The term $H_{stat}$ is strongly dependent on the thickness of the retentive layer. For this discussion it is supposed that the retentive layer is a monomolecular layer of a bonded silane ($C_{18}$) to the FS. In this case also $H_{stat}$ can be neglected. All kinetic effects are accounted in the contribution $H_{kin}$.

Therefore equation (5.7) reduces to

$$H = H_{diff} + H_{trans} + H_{kin} \qquad (5.8)$$

The only factor in this equation that will be influenced by the flow velocity profile is $H_{trans}$. The expression for $H_{trans}$ therefore will be different for pressure-driven and electrically driven flow.[56]

$$H_{t,lam} = \frac{(11k^2 + 6k + 1)}{(1+k)^2} \frac{r_c^2 u}{24 D_i} \qquad (5.9)$$

$H_{t,lam}$ is the contribution to plate height from resistance to mass transfer for pressure-driven flow. The same contribution for EOF has been already given in equation (5.5). Thus, two HETP equations can be given (equations 5.6 and 5.10), for OTLC and OTCEC. Because the contribution to HETP by the injection zone width is also negligible, this term was eliminated and the following two equations were used to calculate $H$ *versus* $u$ curves for OTCEC and OTLC:

$$H = \frac{2D_i}{u} + \frac{uk^2}{(1+k)^2} \frac{r_c^2}{4D_i} + \frac{2uk}{(1+k)^2 k_d} \tag{5.10}$$

$$H = \frac{2D_i}{u} + \frac{(11k^2 + 6k + 1)}{(1+k)^2} \frac{r_c^2 u}{24D_i} + \frac{2uk}{(1+k)^2 k_d} \tag{5.11}$$

The results are given in Figure 5.2.

The minimum value of the $H$ *versus* $u$ curve is similar in both cases. For example for a solute with $k = 2$, the HETP minimum decreases by *ca.* 33% from 7.5 $\mu$m in OTLC to 5 $\mu$m in OTCEC. More pronounced though is the HETP improvement at higher velocities. For example the same solute at a velocity of 5 mm s$^{-1}$ will have an HETP value of 24 $\mu$m in OTLC and 10 $\mu$m in OTCEC. With that observation in mind, one may consider the use of wider i.d. capillaries for OTCEC. This will definitely improve UV–Vis detection because of a longer path-length. Therefore in Figure 5.3, the $H$ *versus* $u$ curve in OTCEC mode is calculated for the case where the capillary is doubled to 20 $\mu$m. All other conditions were kept the same.

Comparing the $H$ *versus* $u$ curves in Figures 5.3 and 5.2(a) leads to the interesting observation that the two curves are very similar. Therefore, one may state that owing to the more favorable flow velocity distribution in the electro-driven case compared with the pressure-driven one – plug flow *versus* parabolic flow velocity profile – the capillary i.d. for OTCEC can be increased by a factor of almost two without severe penalty in HETP and therefore plate number. The benefits of EOF for capillary LC become most prominent at high velocities. The

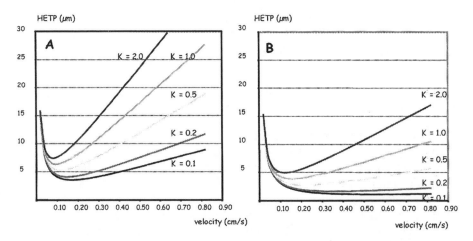

**Figure 5.2** *Calculated* H *versus* u *curves (A) according to equation (5.11) for OTLC and (B) from equation (5.10) for OTCEC. Diffusion coefficient, 1.5 × 10$^{-5}$ cm$^2$ s$^{-1}$; rate constant, 2500 s$^{-1}$; capillary i.d., 10 $\mu$m*

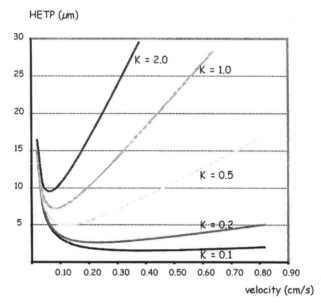

**Figure 5.3**  *Calculated* H *versus* u *curve for OTCEC with a capillary i.d. of 20 μm. All other conditions and parameters as for Figure 5.2*

question though is, what velocity can be achieved in OTCEC under the constraint that a maximum voltage of 30 kV is available in practice?

The voltage that is needed to reach a certain velocity in a capillary column of length $L$ is given by:

$$V = \frac{uL\eta}{\varepsilon_0 \varepsilon_r \zeta} \qquad (5.12)$$

and for a pressure-driven case the pressure is given by

$$\Delta P = \frac{32ul\eta}{d_c^2} \qquad (5.13)$$

These calculations were done for the same capillary column as used in Figure 5.2. Therefore the HETP values for this capillary column in the pressure- and electro-driven mode at the particular velocity are also available. These data are summarized in Figure 5.4.

A zeta potential of 35 mV was used in the calculation. It was assumed that the fused silica capillary has a mono-molecular $C_{18}$ silane layer bonded to its surface. The value of the zeta potential is close to experimental values measured by Dittmann and Rozing for silica-based RP column[56] and lower than the 100 mV that one may expect for a bare fused silica tube.

Figure 5.4 shows again that at the same velocity the capillary column in

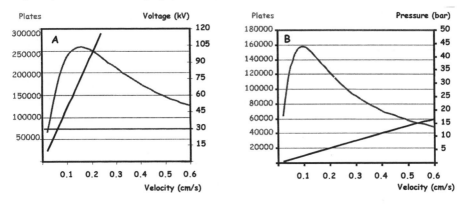

**Figure 5.4** *Calculation of (A) the voltage or (B) the pressure required to drive the mobile phase, and the plate number obtained. k = 1; diffusion coefficient $D_i$ = 2500 $s^{-1}$; viscosity $\eta$ = 0·87 cP; capillary length L = 1 m; capillary i.d. = 10 μm, zeta potential $\zeta$ = 35 mV*

electro-driven mode, delivers 2–2.5 times more plates. On the other hand, the figure also shows that the maximum speed that can be obtained in electro-driven mode under these conditions is 0.06 cm s$^{-1}$. In the pressure-driven mode, though, only 16 bar is required to drive the mobile phase through the column. So in the pressure-driven mode one has the option to increase the length of the column and the velocity of the mobile phase by increasing the pressure and thus obtain the same number of plates in the same time in OTLC as in OTCEC.

This has been calculated for the following example. Suppose one wants to obtain 200 000 plates in a 10 μm i.d. open capillary coated with an ODS layer. The solvent contains 70% acetonitrile and the solute has $k$ = 0.1. The values for all other parameters used to calculate the HETP value at a particular velocity are as given in Figures 5.2–5.4. In this calculation the column is elongated so that, with the desired velocity and the resulting HETP value, 200 000 plates are achieved. Further the voltage and pressure to drive the solvent through the column are calculated as well as the time it takes for the solute with $k$ = 0.1 to elute. The results of these calculations are given in Figure 5.5.

A clear observation can be made. Under these conditions in OTCEC mode, one obtains the required plate number, in the following way. In order to generate 200 000 plates per column, for example, at 0.2 cm s$^{-1}$, a capillary length of 34 cm is required. Under these hypothetical conditions, one needs 30 kV to obtain this velocity. A migration time of 3.1 min is predicted for the solute with $k$ = 0.1 [see right hand $y$-axis in Figure 5.5(a)].

In the OTLC mode of operation one needs a capillary length of *ca.* 72 cm to obtain the required plate number but only a pressure of *ca.* 31 bar to obtain the velocity of 0.2 cm s$^{-1}$. The migration time for the solute with $k$ = 0.1 doubles to 6.6 min.

However, a capacity ratio 0.1 is not a very practical value in liquid chromatography. The next example uses a capacity ratio 1.0. As was seen

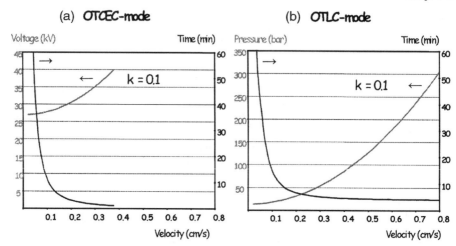

**Figure 5.5**    (a) *Plot of voltage required in OCTEC and* (b) *plot of pressure required in OTCL to obtain 200 000 plates, and analysis time for a solute with* k = 1 *versus velocity of the solvent. Capillary i.d. – 10 µm; solvent is 70% acetonitrile in dilute buffer. Other parameters as given in Figures 5.2–5.4*

before in Figures 5.2 and 5.3, the HETP values quickly increase owing to the growing contribution of trans-channel diffusion to the HETP. In Figure 5.6 $k = 1$ has been inserted in the equations.

As seen before, with an increase in $k$ value, the HETP values also increase. As a consequence one needs a longer capillary column to achieve the required plate number, in this case 200 000. Now the 30 kV constraint becomes an obstruction for OTCEC. With maximum applied voltage 30 kV, the field strength in the long capillary decreases and a low velocity is obtained. For example, at 30 kV a velocity of approx. 0.05 cm s$^{-1}$ occurs in a capillary with sufficient length to generate 200 000 plates. The migration time for the solute with $k = 1$ will be close to two hours.

On the other hand, in OTLC mode, one can operate the capillary column at 400 bar. With a capillary length of 357 cm, a velocity of 0.5 cm s$^{-1}$ can be achieved with the required plate number. The migration time of a solute with $k = 1$ will be *ca.* 23 min under these conditions.

The intrinsically higher efficiency of open tubular liquid chromatography in the electro-driven mode will allow using shorter capillaries to obtain the required efficiency in shorter time than in OTLC at low $k$ values. On the other hand, the OTCEC system is constrained by the availability of the maximum voltage that can be delivered (30 kV in commercial CEC instruments). Further lengthening of the column will lead to a reduction in velocity and therefore increase of migration time.

This treatment demonstrates that because of the more favorable flow velocity distribution in capillary columns where the solvent movement is propagated by electrical field force, the overall HETP value does indeed decrease. However,

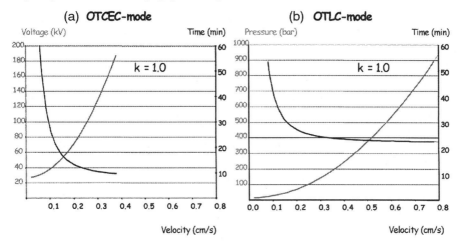

**Figure 5.6** *Plots as for Figure 5.5, but with* k = 1

owing to the 30 kV constraint in practical and commercial CEC equipment, this higher efficiency can be exploited only in relatively short capillaries and at low *k* values. In this case, the analysis time will be consequently very short. So one may argue that OTCEC will be eventually best suited for high-speed, high-efficiency capillary separations. In this mode the capillary i.d. can even be doubled compared with a pressure-driven mode. Because a high electrical field is more easily achieved in planar, microfluidic separation systems, one may anticipate that this field will eventually benefit.

OTLC on the other hand will eventually be a technique equivalent to capillary GC when long capillary columns at high mobile phase velocity are used. Decreasing the capillary i.d. will favor the OTLC mode.

It should, however, be kept in mind that this treatment applies to solutes that are uncharged and separate by partition chromatograph. When the solutes are charged, their electrophoretic mobility will add another component to their overall velocity. Also, the inevitable electrophoretic contributions to zone broadening affect the above treatment. It is easy to predict that systems that combine both modes of driving the solvent through a separation capillary or microfluidic channel will lead to the most powerful separation devices. The high separation efficiency and additional manipulation of selectivity by the application of the electric field will extend the capabilities of such devices beyond what is currently achievable.

## 4 Improving the Phase Ratio for OTCEC

The low phase ratio of the column, besides the low sensitivity of UV–Vis absorption detection, remains the main problem in OT capillary LC separation methods and leads to low sample capacity and loadability. This problem has been described and discussed by several authors in conjunction with OTLC.

Poppe and co-workers have addressed the issue by using thick layer polymeric phases on the capillary wall.[57–59] The drawback of such a phase, however, is the low diffusion coefficient of the solute in the retentive layer. This leads to a significant contribution to the overall HETP for retained solutes (just the term that was eliminated in the discussion at the beginning of Section 3). However it is questionable whether such a capillary column will have the surface charge necessary to generate a zeta potential and therefore an EOF.

To date, three different approaches have been published to increase the phase ratio of a capillary column and at the same time maintain an EOF. These approaches will be discussed here.

The phase ratio $\beta$ of an open tubular capillary column can be estimated from:

$$\beta = \frac{r_c^2}{(r_c - \delta)^2} - 1 \qquad (5.14)$$

in which $\delta$ is the stationary phase film thickness. For a 10 $\mu$m i.d. OT capillary column with a coating that is a monomolecular layer of an alkylsilane (2.5 nm) a phase ratio of 0.001 is calculated. For a packed column, a phase ratio of 0.1 is typical. If a thick stationary phase layer of say 0.1 $\mu$m is deposited on the surface of the capillary the phase ratio increases to 0.04. However, as mentioned before, such thick retentive layers will contribute significantly to band spreading because of the low diffusivity of the solutes in the layer.

Colon and co-workers[60,61] deposited a copolymer of tetraethoxysilane (TEOS) and *n*-octyltriethoxysilane ($C_8$-TEOS) on the inner surface of 13 $\mu$m i.d. fused silica capillaries. The copolymer was generated by acid hydrolysis of a mixture of TEOS/$C_8$-TEOS by which a sol (a colloidal suspension of very small particles) is formed. The sol gelates and is then transferred into the capillary. After a short time the gel is forced out of the capillary, leaving a film of TEOS/$C_8$-TEOS copolymer on the surface of the capillary. Overnight thermal treatment at 120 °C hardens the hydrogel and leaves a $C_8$-modified sub-micron particle silica glass layer on the capillary surface.

An illustration that a truly RP-type retentive layer has been obtained is given in Figure 5.7. Panel A shows the separation obtained for the solutes in a neutral test mixture on a capillary prepared by this method with 13 $\mu$m i.d. and a length of 50 cm. Conditions typical for reversed-phase chromatography were applied *i.e.* methanol/aqueous buffer 2/1. Panel B shows the result obtained under the same conditions with a capillary coated with TEOS. The absence of separation demonstrates the effect of the retentive layer in the coated capillary.

The EOF obtained in this case is *ca.* 0.1 cm s$^{-1}$ at 30 kV, which is significantly lower than predicted from Figure 5.4. Therefore one must suppose that under these conditions the interface had a low surface charge. (The pH of the buffer was not specified by these authors.) This also confirms the theoretical finding in Section 3 that the available voltage quickly becomes the limiting factor for speed in OTCEC.

The layer thickness of these capillaries and therefore the phase ratio can be

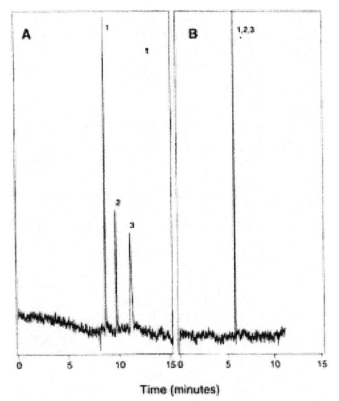

**Figure 5.7** *Separation of a text mixture containing three polyaromatic hydrocarbon (PAH) compounds by OTCEC in two different capillaries:* (A) *capillary coated with copolymer $C_8$-TEOS/TEOS, ratio 0·2;* (B) *capillary coated with TEOS only. Separation conditions: column i.d. 13 µm, length 50 cm (to detector window); mobile phase methanol/1 mM phosphate buffer 2/1, voltage 30 kV; detection 220 nm. Peak identification: 1, naphthalene; 2, phenanthrene; 3, pyrene*
(Reproduced from Y. Guo and L.A. Colon[60] with permission of the publishers of *Anal. Chem.*)

influenced by variation of the ratio TEOS/$C_8$-TEOS. A linear increase of the capacity ratio of neutral solutes is found when this ratio increases. In practice the increase of retention can be offset by increasing the proportion of the organic solvent in the mobile phase without affecting the efficiency.

A clear demonstration of the effect of the increased phase ratio is given in Figure 5.8. Here the separation of a mixture of five PAH compounds was compared using a capillary column coated *via* the sol–gel process and a capillary column where a monomolecular layer of *n*-octyldimethylsilane was coated in a conventional manner. Under similar conditions to Figure 5.7 the capacity ratios of the solutes were 4–5 times higher with the sol–gel coated capillary (panel A) than with the conventional coated capillary (panel B) even

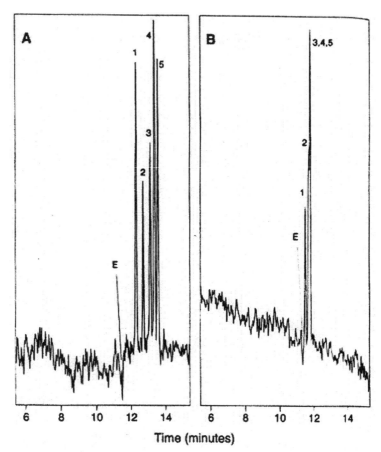

**Figure 5.8** *Electrochromatograms of a text mixture of five PAHs.* (A) *Capillary coated by sol–gel method (C₈-TEOS/TEOS ratio 0·4), mobile phase ratio (methanol/buffer) 70/30;* (B) *capillary coated with* n-*octyldimethyloctylchlorosilane, mobile phase ratio (methanol/buffer) 60/40. Separation conditions are given in Figure 5.7. Peak identification: 1, naphthalene; 2, biphenyl; 3, fluorene; 4, 2-ethylnaphthalene; 5, 2,6-dimethylnaphthalene. E indicates electroosmotic mobility*
(Reproduced from Y. Guo and L.A. Colon[60] with permission of the publishers of *Anal. Chem.*)

with a lower proportion of methanol used in the mobile phase to increase retention.

The efficiency of the capillaries prepared in this way agrees with the graph in Figure 5.4(A), which supports the theoretical treatment used.

Capillary columns prepared by the sol–gel method showed good longevity and stability after treatment with low and high pH mobile phases for extended periods. This can be attributed to the fact that the retentive layer is not bonded to the surface *via* siloxane bonds as in the conventional bonding procedures but is incorporated in the bulk of the retentive layer.

Pesek and co-workers took a somewhat different approach.[62-64] This group etched the surface of the capillary by hydrolytic treatment of the bare silica with ammonium hydrogen difluoride. After prior treatment with concentrated hydrochloric acid, the capillary was etched at varying temperatures and durations. These parameters determine the morphology of the surface that is obtained in the etching process. The authors found that conditions leading to a spongy, sand-dune kind of surface obtained in the treatment provided the best compromise between an increase of the phase ratio and the robustness of the surface structures obtained (Figure 5.9). The spongy silica surface obtained in this way was modified to a reversed-phase-type layer by a silanization/hydrosilanization procedure pioneered by this group. In this process, first, almost all silanol groups on the fused silica surface are covered by modification with triethoxysilane. This leaves a silicon hydride layer on the surface, which serves as a hook to connect an alkane group *via* a hydrosilation reaction.

The initial work of this group was done with 50 $\mu$m i.d. fused silica capillaries. An example is given in Figure 5.10. In Figure 5.10(a) the lack of separation of two proteins, turkey (1) and chicken (2) lysozyme on a bare fused silica capillary at a low pH of the run buffer is demonstrated. The low pH was selected to

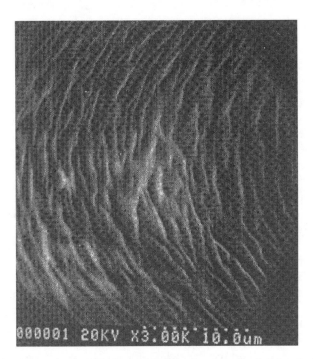

**Figure 5.9** *Scanning electron photograph of the inner wall of a fused silica capillary etched with ammonium hydrogen difluoride at 300 °C for 4 h*
(Reproduced from J.J. Pesek and M.T. Matyska[64] with permission of the publishers of *J. Cap. Electrophoresis*)

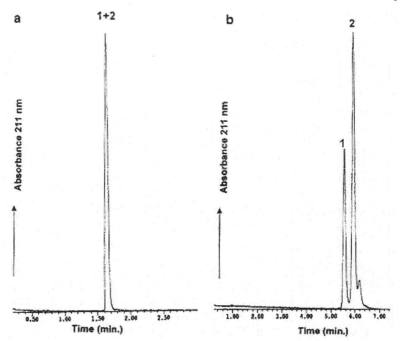

**Figure 5.10** *Separation of turkey* (1) *and chicken* (2) *lysozyme on* (a) *a bare capillary and* (b) *a $C_{18}$-modified capillary. Capillary i.d. = 50 μm, length = 45 cm; run buffer pH = 2.14*
(Reproduced from J.J. Pesek and M.T. Matyska[64] with permission of the publishers of *J. Cap. Electrophoresis*)

suppress the electroosmotic flow so that the selectivity of separation is only related to electrophoretic mobility differences and, in Figure 5.10(b), to chromatographic interactions. The lack of separation in the CE mode is not unexpected because the two proteins differ only in one amino acid. In Figure 5.10(b) the separation was obtained under the same conditions and therefore can only be due to the chromatographic interactions. So in this case, one is dealing with selectivity enhancement of an electrophoretic (CZE)-type separation assisted by chromatographic interaction. As argued in Section 2, the price for retention in CE is loss in efficiency as illustrated in Table 5.1.

The chromatographic interactions of the solute bradykinin with the $C_{18}$-coated layer on the fused silica inner wall decreases the efficiency and peak shape dramatically. If methanol is added to the run buffer, this reduces the chromatographic interactions by increased solvating power, and increases the efficiency and improves the peak symmetry (last line of Table 5.1).

The EOF velocity found for the etched, $C_{18}$-modified capillary was very low, and, as in packed column CEC, depends on the mobile phase composition. Figure 5.11 shows the EOF velocity dependence on the proportion of methanol in the run buffer. The plot closely follows the expected track based on the

**Table 5.1** *Efficiency and peak symmetry for bradykinin on different capillaries*

| Capillary | Buffer pH | Plate number | Symmetry |
|---|---|---|---|
| Bare fused silica | 3.7 | 400 000 | 1.00 |
| Etched, 300 °C, 3 h | 3.7 | 337 000 | 0.73 |
| Etched, Si–H modified | 3.7 | 542 000 | 0.73 |
| Etched, $C_{18}$ modified | 3.7 | 68 000 | 1.12 |
| Etched, $C_{18}$ modified | 3.0 | 33 600 | 2.47 |
| Etched, $C_{18}$ modified | 3.0 with 10% methanol | 79 000 | 2.22 |

Efficiency calculated from the width at half peak height. Symmetry calculated at 10% of peak height. Buffer lactic acid/$\beta$-alanine pH 3.7. Voltage 25 kV. Capillary length 50 cm (to the detector window), overall length 70 cm, i.d. 50 $\mu$m.

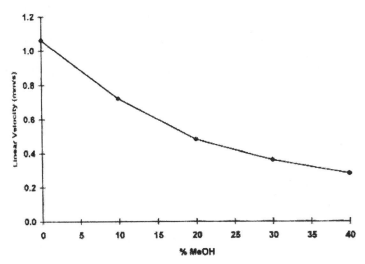

**Figure 5.11** *EOF velocity measured with a neutral marker, dimethyl sulfoxide (DMSO), in an etched, $C_{18}$-modified capillary as a function of the proportion of methanol in the run buffer at pH 2.14*
(Reproduced from J.J. Pesek and M.T. Matyska[62] with permission of the publishers of *J. Chromatogr.*)

change in the quotient of the dielectric constant and the viscosity as reported for packed column CEC.[47]

Nevertheless, this example, and others provided by this group cannot be regarded as an OTCEC experiment but rather as the mode of separation described in Section 2, *i.e.* CE with chromatographic interaction. Because of the large internal diameter of the capillaries used, 50 $\mu$m, the phase ratio for chromatography becomes very small. With the same assumptions as at the beginning of this section, phase ratios of 0.0002 and 0.008 are calculated from equation (5.14) for a retentive layer of 2.5 and 100 nm respectively. In addition because of the very low value or absence of EOF there is barely any flow-driven transport of the solutes through the retentive tube, which is characteristic for chromatographic separation.

In another paper the authors describe the same capillary modification technique applied to much narrower i.d. fused silica capillaries.[65] In this work fused silica capillaries with 20 $\mu$m i.d. were used. Quite remarkably, at low values of the pH of the solvent, an anode EOF was observed. This finding indicates the presence of positive charges on the surface at low pH. The authors postulated the presence of ammonium sites responsible for the flow reversal. In practice though this means again that in such kinds of capillaries there will be very low EOF, and therefore a low flow transport contribution for the solutes.

The i.d., lower than that used in the previous work, should however improve the efficiency of separation. The separation of the two lysozymes from turkey and chicken performed under the same conditions as in Figure 5.10 led to the chromatogram in Figure 5.12. A clear improvement of efficiency is indeed observed.

Although, because of the very low (or absent) flow transport, this approach in a pure sense may not be regarded a OTCEC, it is a very promising approach to separations. In particular it may become the method of choice for the separation

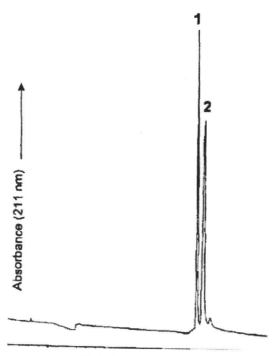

**Figure 5.12**   *Separation of turkey* (1) *and chicken* (2) *lysozyme on a 20 $\mu$m i.d. etched, $C_{18}$-modified fused silica capillary. Voltage 30 kV; capillary length 51.5 cm; other conditions as in Figure 5.10*
(Reproduced from J.J. Pesek and M.T. Matyska[65] with permission of the publishers of *J. Chromatogr.*)

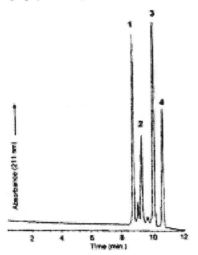

**Figure 5.13** *Separation of a mixture of cytochrome c's on 20 μm i.d. etched C₁₈-modified fused silica capillary. Voltage 15 kV; other conditions as for Figure 5.12. Peak identification: 1, horse; 2, bovine; 3, chicken; 4, tuna*
(Reproduced from J.J. Pesek and M.T. Matyska[65] with permission of the publishers of *J. Chromatogr.*)

of proteins. This is well illustrated by the chromatogram in Figure 5.13, which shows the separation of 4 cytochrome c's on an etched, $C_{18}$-modified capillary. Conditions are similar to those in Figure 5.12, though the voltage applied was reduced to 15 kV.

From the EOF *versus* pH plot one would expect an anodic flow at the pH 3.7 that is used in this experiment. So flow transport present will actually oppose the direction of the solutes. This may be a very attractive property of this system, as it keeps the solutes longer in the separation capillary and therefore allows more time for separation.

Remcho and co-workers[66,67] used an approach originally reported by Poppe and co-workers[57-59] to prepare thick layers of polymethacrylate on the inside of a fused silica capillary in an attempt to increase the phase ratio of a capillary column. After hydroxylation of the surface by acid treatment, 3-trimethoxysilylpropyl methacrylate is bonded to the capillary wall. The vinyl group of the acrylate provides a chemical hook for subsequent polymerization with *n*-butyl methacrylate or with a mixture of *n*-butyl methacrylate with 1,4-butanediol methacrylate. The last monomer functions as a cross-linker. In this way, a layer of linear or cross-linked polyacrylate was obtained. This layer was carefully dried and cured at 120 °C in order to obtain a stable film of polyacrylate on the inside of the capillary serving as a reversed-phase type retentive layer.

Varying the concentration of monomer and cross-linker influences the retentive properties of this layer. Increase of the monomer concentration leads to higher capacity factors, although beyond a certain concentration of the

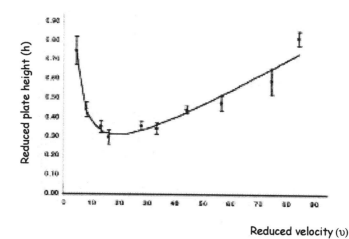

**Figure 5.14** *Effect of flow velocity on plate height for an unretained solute (acetone). Capillary with a linear polyacrylate layer prepared with 45% monomer (n-butyl acrylate) concentration; i.d. 25 μm, length to detector 20 cm* (Reproduced from Z.J. Tan and V.T. Remcho[66] with permission of the publishers of *Anal. Chem.*)

modifier retention on linear polyacrylate films started to decrease. The same observation was made when using cross-linked polyacrylate.

Unfortunately, these capillaries also show a relatively low EOF. Values range between 0.1 and 0.2 cm s$^{-1}$ at 30 kV on a 45 cm capillary with a mobile phase of phosphate buffer (pH 7) acetonitrile 4/1.

The work was done with 25 μm i.d. capillaries. A reduced parameter plot ($h/v$ curve or Knox plot) was obtained and is given in Figure 5.14.

A minimum reduced plate height of 0.3 was found which is an HETP value of 8 μm which is higher than our prediction in Figure 5.3 (although conditions are not very comparable).

The cross-linked phases appeared to have better efficiency for the unretained solute. Plate heights of 3–4 μm were obtained in a 25 μm i.d. capillary with a cross-linked layer of *n*-butyl methacrylate and 1,4-butanediol dimethacrylate. This is in agreement with the predictions of Figure 5.3 for a 20 μm i.d. capillary. However, as expected, for retained solutes the efficiency decreases rapidly by a factor of 2–4 depending on the *k* of the retained solute. The *k* itself depends on the monomer/cross-linker concentration in the coating mixture.

In a final experiment, these authors quantitated the effect of pressurized flow on the HETP and indicated the potential of such capillaries when one uses a combination of pressure-driven and electro-driven flow to separate solutes. This is shown in Figure 5.15.

The upper two traces allow a direct comparison of pressure-driven OTLC and electro-driven OTLC. Pressure was set so that the velocity in both cases is the same, illustrated by the equivalence of retention times in both traces. It is clear though that the efficiency of the peaks in the pressure drive case is a factor of 2–3

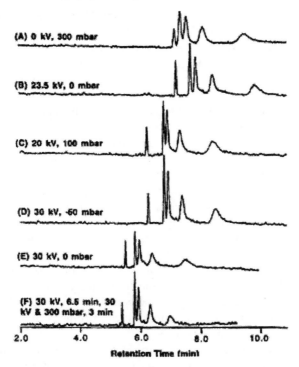

**Figure 5.15** *Comparison of OTCEC and OTLC separations. Capillary: cross-linked poly-(n-butyl methacrylate), total length 60 cm, length to detector 45 cm. Solutes: 1, acetone; 2, methyl paraben; 3, ethyl paraben; 4, n-propyl paraben; 5, n-propyl paraben. Solvent: phosphate buffer (pH 7)/acetonitrile 4/1* (Reproduced from Z.J. Tan and V.T. Remcho[66] with permission of the publishers of *Anal. Chem.*)

worse than in the OTCEC mode. In combining pressure and voltage one can have both transport mechanisms cooperating as in trace 3, and gain in speed, or opposing so that one may obtain a longer time for separation.

Also, as expected from the discussion in Section 3, the pressure required to drive the solvent at a particular speed is very low.

## 5   Conclusions

A few striking conclusions can be derived from the treatment in Sections 2 and 3 and the experimental results described in Section 4.

In OTCEC the flat, plug flow velocity profile generated by electromotive forces has a large impact on zone broadening compared with pressure-driven flow. A significant reduction of the HETP values by a factor of 2–3 for non- or slightly retained solutes is expected theoretically and demonstrated experimentally. As theory predicts, with increase of retention the HETP increases rapidly, particularly when the capillary i.d. is larger than 10 $\mu$m. Nevertheless, one can safely state that because of the more favorable zone broadening in the OTCEC case, one may be able to use a two times larger i.d. capillary than in OTLC, alleviating the poor UV–Vis detection sensitivity caused by the short light path.

On the other hand, the voltage used as a driving force for OTCEC quickly reaches its practical limit of 30 kV when capillaries are lengthened for high efficiency. In this respect, OTLC will easily outperform OTCEC. Electro-driven separations therefore will only be practical with short ($< 50$ cm) capillary columns where high fields are achievable and therefore good velocities are achieved. To achieve very high plate numbers, long capillary columns that run at high velocity in OTLC mode are necessary.

Attempts to improve the phase ratio of capillary columns are penalized by loss of EOF. The attempts described in this chapter all lead to values $< 0.2$ cm s$^{-1}$. Different coatings may be required.

Finally, a combination of OTLC and OTCEC seems the way to go. In particular because low pressure, $< 10$ bar, is sufficient to obtain a speed of a few millimetres per second in a capillary of 10 $\mu$m i.d. Because this capability is available in modern CE instruments, the technique can be exploited without large technical and practical hurdles. The combination of separation mechanisms, namely partitioning and electromigration, promises an excellent way to manipulate selectivity of separation.

Charged solutes will add an electrophoretic velocity component to their transport mechanism. The combination of pressure-driven and electro-driven transport with the solute's electrophoretic migration, in an appropriate way, will provide a very powerful way to tune the selectivity of separation. Practical models to predict retention in such systems will be required for this method to be used in practice. Planar, microfluidic channels seem predestined for such systems. One can expect a large interest in this field in the near future.

## Acknowledgement

The authors want to thank their colleague Gordon Ross of the Hewlett-Packard Waldbronn Analytical Division, for helpful comments and corrections.

## References

1. J.H. Knox and I.H. Grant, *Chromatographia*, 1987, **24**, 135.
2. M.M. Dittmann, F. Bek, K. Wienand, and G.P. Rozing, *LC-GC*, 1995, **13**, 800.
3. J.W. Jorgenson and K.D. Lukacs, *J. Chromatogr.*, 1981, **218**, 208.
4. J.H. Knox and I.H. Grant, *Chromatographia*, 1991, **32**, 317.
5. H. Yamamoto, H. Baumann, and F. Erni, *J. Chromatogr.*, 1992, **593**, 313.
6. C. Yan, D. Schaufelberger, and F. Erni, *J. Chromatogr.*, 1994, **670**, 15.
7. T. Tsuda, *LC-GC*, 1992, **5**, 26.
8. S. Kitagawa and T. Tsuda, *J. Microcol. Sep.*, 1994, **6**, 91.
9. S. Kitagawa and T. Tsuda, *J. Microcol. Sep.*, 1995, **7**, 59.
10. N.W. Smith and M.B. Evans, *Chromatographia*, 1994, **38**, 649.
11. N.W. Smith and M.B. Evans, *Chromatographia*, 1995, **41**, 197.

12. R.J. Boughtflower *et al.*, *Chromatographia*, 1995, **41**, 398.
13. R.J. Boughtflower *et al.*, *Chromatographia*, 1995, **40**, 329.
14. B. Behnke and E. Bayer, *J. Chromatogr.*, 1994, **680**, 93.
15. B. Behnke, E. Grom, and E. Bayer, *J. Chromatogr.*, 1995, **716**, 207.
16. H. Rebscher and U. Pyell, *Chromatographia*, 1994, **38**, 737.
17. T. Eimer, K.K. Unger, and T. Tsuda, *Fresenius' J. Anal. Chem.*, 1995, **352**, 649.
18. J.L. Liao, N. Chen, C. Ericson, and S. Hjerten, *Anal. Chem.*, 1996, **68**, 3468.
19. C. Yan, R. Dadoo, R.N. Zare, D.J. Rakestraw, and D.S. Anex, *Anal. Chem.*, 1996, **68**, 2726.
20. A.S. Rathmore and C. Horvath, *J. Chromatogr. A*, 1996, **743**, 231.
21. M.R. Euerby, C.M. Johnson, K.D. Bartle, P. Myers, and S.C.P. Roulin, *Anal. Commun.*, 1996, **33**, 403.
22. H. Rebscher and U. Pyell, *J. Chromatogr. A*, 1996, **737**, 171.
23. M.M. Robson, S. Roulin, S.M. Shariff, M.W. Raynor, K.D. Bartle, A.A. Clifford, P. Myers, M.R. Euerby, and C.M. Johnson, *Chromatographia*, 1996, **43**, 313.
24. M.T. Dulay, C. Yan, D.J. Rakestraw, and R.N. Zare, *J. Chromatogr. A*, 1996, **725**, 361.
25. F. Lelievre, C. Yan, R.N. Zare, and P. Gareil, *J. Chromatogr. A*, 1996, **723**, 145.
26. C. Fujimoto, Y. Fujise, and E. Matsuzawa, *Anal. Chem.*, 1996, **68**, 2753.
27. H. Minakuchi, K. Nakanishi, N. Soga, N. Ishizuka, and N. Tanaka, *Anal. Chem.*, 1996, **68**, 3498.
28. J.-L. Liao, N. Chen, C. Ericson, and S. Hjerten, *Anal. Chem.*, 1996, **68**, 3468.
29. M.M. Dittmann and G.P. Rozing, *J. Chromatogr. A*, 1996, **744**, 63.
30. S.E. van den Bosch, S. Heemstra, J.C. Kraak, and H. Poppe, *J. Chromatogr. A*, 1996, **755**, 165.
31. M.R. Taylor, P. Teale, S.A. Westwood, and D. Perrett, *Anal. Chem.*, 1997, **69**, 2554.
32. A. Maruska and U. Pyell, *Chromatographia*, 1997, **45**, 229.
33. S. Nilsson, L. Schweitz, and M. Petersson, *Electrophoresis*, 1997, **18**, 884.
34. M.R. Taylor and P. Teale, *J. Chromatogr. A*, 1997, **768**, 89.
35. L. Schweitz, L.I. Andersson, and S. Nilsson, *Anal. Chem.*, 1997, **69**, 1179.
36. T.M. Zimina, R.M. Smith, and P. Myers, *J. Chromatogr. A*, 1997, **758**, 191.
37. J.M. Lin, T. Nakagama, X.Z. Wu, K. Uchiyama, and T. Hobo, *Fresenius' J. Anal. Chem.*, 1997, **357**, 130.
38. C. Wolf, P.L. Spence, W.H. Pirkle, E.M. Derrico, D.M. Cavender, and G.P. Rozing, *J. Chromatogr. A*, 1997, **782**, 175.
39. M.M. Robson, M.G. Cikalo, P. Myers, M.R. Euerby, and K.D. Bartle, *J. Microcol. Sep.*, 1997, **5**, 357.
40. M.R. Euerby, D. Gilligan, C.M. Johnsson, S.C. Roulin, P. Myers, and K.D. Bartle, *J. Microcol. Sep.*, 1997, **5**, 373.
41. P. Sandra, A. Dermaux, V. Ferraz, M.M. Dittmann, and G.P. Rozing, *J. Microcol. Sep.*, 1997, **5**, 409.
42. J.M. Ding and P. Vouros, *Anal. Chem.*, 1997, **69**, 379.
43. J.H. Miyawa, M.S. Alasandro, and C.M. Riley, *J. Chromatogr. A*, 1997, **769**, 145.
44. G.A. Lord, D.B. Gordon, P. Myers, and B.W. King, *J. Chromatogr. A*, 1997, **768**, 9.
45. C. Ericson, J.-L. Liao, K. Nakazato, and S. Hjerten, *J. Chromatogr. A*, 1997, **767**, 33.
46. M.M. Dittmann and G.P. Rozing, *J. Microcol. Sep.*, 1997, **5**, 399.
47. K.D. Altria, N.W. Smith, and C.H. Turnbull, *Chromatographia*, 1997, **46**, 664.
48. A. Palm and M. Novotny, *Anal. Chem.*, 1997, **69**, 4499.
49. R.M. Seifar, W.Th. Kok, J.C. Kraak, and H. Poppe, *Chromatographia*, 1997, **46**, 131.
50. J.H. Knox and M.T. Gilbert, *J. Chromatogr.*, 1979, **186**, 405.
51. J.H. Knox, *Chromatographia*, 1980, **18**, 453.
52. J.W. Jorgenson and K. Lukacs, *Anal. Chem.*, 1981, **53**, 1298.
53. S. Swedberg and D. McManigill in 'Techniques in Protein Chemistry', ed. T. Hugli, Academic Press, 1989, p. 472.

54. G.P. Rozing, *Am. Lab.*, December 1998, 33.
55. J.C. Giddings in 'Dynamic of Chromatography, Part 1, Principles and Theory', Marcel Dekker, New York, 1965.
56. M.M. Dittmann and G.P. Rozing, *J. Microcol. Sep.*, 1997, **5**, 399.
57. R. Swart, J.C. Kraak, and H. Poppe, *J. Chromatogr.*, 1994, **670**, 25.
58. R. Swart, J.C. Kraak, and H. Poppe, *Chromatographia*, 1995, **40**, 587.
59. R. Swart, J.C. Kraak, and H. Poppe, *J. Chromatogr.*, 1995, **689**, 177.
60. Y. Guo and L.A. Colon, *J. Microcol. Sep.*, 1995, **7**, 485.
61. Y. Guo and L.A. Colon, *Anal. Chem.*, 1995, **67**, 2511.
62. J.J. Pesek and M.T. Matyska, *J. Chromatogr.*, 1996, **736**, 255.
63. J.J. Pesek, M.T. Matyska, and L. Mauskar, *J. Chromatogr.*, 1997, **763**, 307.
64. J.J. Pesek and M.T. Matyska, *J. Cap. Electrophoresis*, 1997, **5**, 213.
65. J.J. Pesek and M.T. Matyska, *J. Chromatogr.*, in press.
66. Z.J. Tan and V.T. Remcho, *Anal. Chem.*, 1997, **69**, 581.
67. Z.J. Tan and V.T. Remcho, *J. Microcol. Sep.*, 1998, **10**, 99.

CHAPTER 6

# Capillary Electrochromatography (CEC)/Mass Spectrometry (MS)

GWYN A. LORD AND DEREK B. GORDON

## 1 Introduction

Detection in CEC is primarily by UV–Vis absorbance through the unpacked (usually) section of capillary, after removal of a section of the polyimide coating. The main problem with this is the short light path-length, resulting in relatively poor sensitivity, although longer path-length 'bubble' and 'Z' cells have been developed. Fluorescence detection offers increases in sensitivity, providing that the analyte is suitable. MS is arguably the best chromatographic detector available for both qualitative and quantitative analysis; it is highly sensitive with unsurpassed specificity and extends the power of both techniques in a synergistic way. CEC is a good alternative to both conventional HPLC and to capillary electrophoresis (CE), when linked to MS. The higher chromatographic efficiency of CEC enhances sensitivity, enabling MS analysis of very low quantities of sample, especially of biological compounds. Advantages over CE coupled to MS include the ability to apply more sample onto a packed capillary and for the analysis of neutral compounds. The CE technique of micellar electrokinetic capillary chromatography, used for neutral compounds, relies on MS-incompatible quantities of surfactants for micelle formation.

What follows is an introduction to MS with descriptions of the most relevant techniques for coupling with CEC and then discussion of the interfaces that are used.

## 2 Mass Spectrometry

MS offers the facility to identify and characterise chemical entities by enabling the determination of mass-to-charge, $m/z$, ratios of both molecular species and fragments obtained by chemical decomposition of those species. The pattern of peaks represented by particular $m/z$ ratios constitutes the mass spectrum. The value of $z$ must be an integer as it is the number of charges which a particular ion carries. Clearly, when $z = 1$ then the $m/z$ ratio is equivalent to the mass of the ion concerned. Values of $z$ greater than 1 may be achieved which, in appropriate circumstances, enables masses which exceed the upper limit of the instrument to be accommodated by reducing the magnitude of $m/z$.

Although mass is an intrinsic property and is universal, it is not unique and this is exemplified by isomers and similar isobaric entities (at low resolution). Despite this apparent shortcoming, MS offers unprecedented sensitivity as an analytical technique. Where a precursor molecular entity decomposes in the mass spectrometer ion source the presence of a peak in the mass spectrum, at a particular $m/z$ value, and representing a fragment or product ion, depends upon the thermodynamic stability of that fragment. Less-stable fragments may decompose yet further and be not well represented as a proportion of the overall 'total ion current' which is the sum of all the individual ion currents in the mixture of ions. In fact the molecular ion itself may be so unstable, in certain instances, that it decomposes completely and no peak is seen. Careful inter-pretation of the mass spectrum of the precursor and product ions enables the chemical structure of the original analyte material to be elucidated. In fact, despite the isobaric nature of isomers, it is sometimes possible to distinguish between these as different fragmentation pathways result owing to different structural orientations.

Figure 6.1(a) shows a block diagram of a generalized mass spectrometer, which consists essentially of four main parts:

(a) an inlet system for the introduction of sample;
(b) an ionization chamber or ion source in which the sample is converted into ions in the gas phase;
(c) a mass separator or analyser in which the ions are separated by virtue of their $m/z$ ratio and, for high resolution work, an energy analyser; and
(d) a detector system.

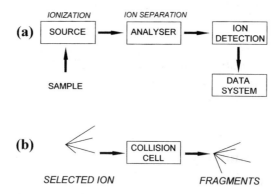

**Figure 6.1.**   (a) *Schematic block diagram of generalized mass spectrometer*. (b) *Schematic diagram of MS/MS*

The computer may, in modern instrumentation, be considered as a fifth part. This latter device in the more sophisticated instruments fulfils the roles of data processing/manipulation and also instrument control. The relative cheapness and ease of use of modern microcomputers enables an operator with adequate keyboard skills to 'drive' very sophisticated mass spectrometers.

## Ionization Methods

The production of ions is a prerequisite to mass analysis. Ions, being charged particles, can be manipulated in a variety of ways but in particular they can be accelerated to high velocity in electric fields and constrained to circular trajectories in magnetic fields.

Furthermore, detector systems may be devised which amplify the current/voltage generated when the separated ion impacts on a sensitive surface. As it is relatively simple to generate mass spectra from either positive or negative ions (but not to analyse both at the same time), it is necessary to consider both cation and anion production. Generally, negative-ion mass spectra are less intense than the corresponding positive-ion spectra, except where the nature of the chemical substance being analysed leads more readily to the formation of stable anions. The more common methods of ionization involve the addition or removal of electrons or protons. In the latter case a so-called pseudomolecular ion is produced, designated as a protonated or de-protonated molecule ion, which differs by $\pm 1$ from the native molecular mass. It is also possible for cationization by alkali metal ions (Na, K) to occur, and this can sometimes aid interpretation.

The advent of fast atom bombardment (FAB) ionization, invented by Barber *et al.* in 1981,[1,2] stimulated a very rapid expansion of biological MS. A secondary effect was the stimulus for the development of electrospray ionization. Both of these ionization processes allow the introduction of the sample in an aqueous or water miscible medium. Furthermore they lend themselves to the development of interfaced flow systems allowing the coupling or 'hyphenation' or powerful separatory methods thus facilitating the analysis of complex mixtures.

In so called 'static' FAB ionization, the sample, in solution, is added to a suitable liquid matrix such as glycerol. Compatibility between sample solvent and liquid mixture is essential. The admixture is placed on a sample probe which is introduced into the mass spectrometer *via* vacuum locks. A beam of fast atoms is directed at the sample/matrix target and the ions that are generated are accelerated out of the source region, separated in the analyser regions and then detected. One major advantage of the liquid matrix is that the surface is continually replenished, resulting in long sample lifetimes. Dynamic 'continuous flow FAB ionization' (CF-FAB),[3] or 'frit-FAB'[4] demands a modification to the probe. This involves the construction of a hollow or thick-walled tube-like probe, through which a suitable capillary may be passed. The probe tip is now a sintered frit through which liquid may pass. The pre-MS end of the capillary is linked to the separation system. Mechanisms for adding the liquid matrix have to be devised and, in general, the volume flow rate is matched to the rate of evaporation and pumping capability of the mass spectrometer. When this is achieved, a steady-state surface profile, which mimics the situation in the static case, is presented to the atom beam. The flow rate is limited to approximately $5–10 \, \mu l \, min^{-1}$, necessitating a large split ratio and lower sensitivity when analytical-sized HPLC columns are coupled to the technique, hence the advantage in using microbore and capillary columns with lower split ratios or no split.

Electrospray ionization (ESI) again requires input of the sample in solution. In 1968 Dole *et al.*[5] first proposed electrically charged droplets as a source of ions for MS and Fenn *et al.*[6] pioneered its development as a MS interface, leading to the first commercial instrument in 1989. Figure 6.2 shows a schematic representation of an ESI source. Suitable buffers are required and it is essential to avoid non-volatile materials. It must be possible to generate ions in solution so that at the high-voltage spray head fine droplets of solution (atomization) are sprayed into the open source region at atmospheric pressure and carry a number of surface charges of the same polarity. The droplets have a forward velocity which carries them through a series of conically shaped skimmers. A curtain gas, usually dry nitrogen, is passed, transverse travel, between the skimmers and this aids evaporation of the solvent. During this process both the volume of any individual droplet and its surface area are diminishing so that the surface charges are brought closer together. Clearly the force of repulsion increases (inverse square law) and this is counteracted by the surface tension which, at a simple level, may be considered as 'holding the droplet together'. The Rayleigh limit is reached when the coulombic repulsion just overcomes the cohesive effect of the surface tension. Beyond this point the droplet disintegrates into smaller droplets and the process continues until the charge resides on the solute, which then passes into the analyser where the different entities are separated and finally detected.

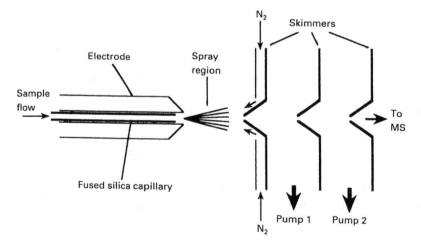

**Figure 6.2**   *Schematic diagram of electrospray source*

The closely related technique of pneumatically-assisted electrospray, or ion-spray, a term instituted by its developers,[7] uses gas, again usually nitrogen, to aid nebulization of the liquid. A development of electrospray uses a corona discharge to ionize sample droplets by chemical ionization, hence it is known as atmospheric pressure chemical ionization (APCI). The technique is ideally suited to non-polar compounds of low molecular mass, but at present has not been coupled with CEC, although there are no reasons to prevent this.

Advantages of electrospray ionization include ease of use, multiple charging of species, thus reducing $m/z$, and a relative lack of the suppression effects observed in FAB ionization although they are not entirely eliminated. For multiple charging to occur, of course, it is necessary for the analyte to possess a sufficient number of sites within its structure to enable gain or loss of the appropriate number of protons.

## Mass Analysers

Analyser systems in mass spectrometry are many and varied. However, for the purposes of the current topic, three types are most appropriate; sector, quadrupole and time-of-flight (TOF) analysers.

Where FAB ionization is used sector analysers are often encountered. The ions are separated according to mass-to-charge ratios, in a magnetic sector designated B. Ions are forced to travel along a particular trajectory, and each ion in turn can be constrained to follow this path, by 'scanning' the magnet either from high to low mass or *vice versa*, usually the former. Each ion in the mixture can therefore be successively brought to focus on the detector. What results from this process is a low-resolution mass spectrum because ions of the same nominal mass may possess slightly different kinetic energies, and therefore momenta, depending on where in the source region they were actually generated and variation in the acceleration to which they have been subjected. Resolution can be dramatically improved by introducing an electrostatic 'energy' analyser, E, which essentially corrects for this variation. This design is the double focussing or two sector instrument. Sector E may precede or succeed sector B, giving rise to forward and reverse geometries allowing different kinds of studies to be performed By including collision cells at appropriate points in the flight path 'collisionally assisted dissociation' (CAD) MS/MS studies are possible (see Figure 6.1b) and multiple sector instruments may be designed.

High costs are involved in the above approach and a cheaper option is the quadrupole analyser. In these analysers ions travel along either stable or unstable trajectories through the inter-rod space. Multiple rod systems (commonly four or six) are used, the rods having both dc and rf voltages applied to opposing conductors (Figure 6.3). The analyser is 'scanned' by changing the magnitude of these voltages whilst keeping the rf/dc ratio fixed. By this means ions of different $m/z$ are successively constrained to stable trajectories and reach the detector.

Multiple mass spectrometry is again possible in a triple quadrupole, or more usually a quadrupole-hexapole-quadrupole arrangement. The initial separation

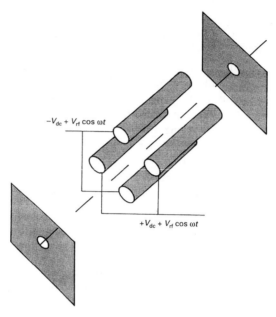

$-V_{dc} + V_{rf} \cos \omega t$

$+V_{dc} + V_{rf} \cos \omega t$

**Figure 6.3** *Schematic diagram of quadrupole analyser*

occurs in the first quadrupole where an ion of interest is selected to undergo CAD in the middle or hexapole collision cell, and the third analyser separates the products of this induced dissociation.

The TOF analyser is probably the simplest of all. In the linear form, ions generated in the source region are allowed to drift along an evacuated tube to the detector. Assuming all ions have the same momentum then clearly mass will be inversely proportional to velocity. Those ions of lower mass will travel faster and arrive at the detector earlier whilst those of increasing mass will travel more slowly and arrive later. These analysers lend themselves to 'pulsed' ionization processes. By recording a start pulse and measuring the time taken for an ion to travel a known distance to the detector and relating this to the mass, the latter may be determined. Clearly variation in the momenta arises and this may be corrected in the more sophisticated reflectron instrument in which an ion with slightly greater momentum (higher velocity) is made to travel along a proportionately longer curved trajectory than an ion with slightly lower momentum (lower velocity). The reflectron functions as an electric mirror, and an error-correcting re-focussing is achieved. Further improvements in the TOF analyser are achieved using time delayed extraction of ions from the source and also examining the phenomenon of post source decay.

One of the most recent developments is a hybrid instrument involving quadrupole and TOF analysers. This is a truly orthogonal method with the TOF constructed at right angles to the quadrupole. Vastly improved sensitivity and mass limit are achieved.

## Detectors in Mass Spectrometry

A wide variety of detectors are available for use in mass spectrometers. Most, however, depend upon the collection of ions, which, when discharged, effect current flow in a conductor. In a suitable circuit the current, or more often an associated voltage, is amplified and measured. Electron multipliers, or photo-multipliers after electron–photon conversion, are commonly used to obtain the high level of amplification required.

There are two main types of chromatographic detectors, *concentration-sensitive*, which respond to changes in analyte concentration, such as UV detectors, where smaller diameter columns show higher sensitivity because the analyte elutes in a smaller volume of eluent and *mass-sensitive*, which respond to changes in the mass flow of analyte.[8] Mass spectrometers, belong to the latter class, in that their response is proportional to the absolute mass of analyte passing through them, rather than to concentration, and they are also analyte destructive. However, an important feature of mass spectrometers fitted with ESI sources, is that they appear to behave in a concentration-sensitive manner, rather than as a mass-sensitive detector.[9] Capillary columns coupled to ESI-MS give higher absolute sensitivity than conventional sized HPLC columns, and the ESI source remains cleaner as much less sample, buffer and contaminants enter the mass spectrometer.

For further reading about mass spectrometry, the reader is encouraged to consult original articles, books and reviews. An excellent review by Burlingame, Boyd and Gaskell[10] is a good starting point.

# 3 Capillary Electrochromatography/Mass Spectrometry Coupling

The first demonstration of coupling an electrochromatographic technique with MS was in 1991 by Verheij *et al.*,[11] who used pseudo-electrochromatography (PEC) and CF-FAB-MS for the analysis of morphine alkaloids and nucleotides. Subsequently, this group coupled PEC with ESI-MS for the analysis of food colours and aromatic glucuronides[12] and more recently sulfonamides and a peptide.[13] The term PEC was used by Verheij *et al.*[11] to distinguish a combination of both pressure-driven and electroosmotic flow (EOF) from pure electrically driven CEC and was originally described by Tsuda in 1987,[14] who used the addition of pressure to EOF to suppress bubble formation. The flow profile in PEC approaches that in pressure-driven chromatography, resulting in an efficiency between those of CEC and HPLC, but biased towards the latter. However, there are advantages in PEC; higher flow rates lead to shorter analysis times and there are no limitations on pH and buffers for generation of EOF, which is not important in PEC.

The combination of purely electrically driven CEC was also first linked with CF-FAB-MS and used for the analysis of steroids,[15] but virtually all subsequent reports have used ESI. A range of compounds have been analysed by CEC/ESI-MS, including textile dyes,[16] peptides,[17] cardiotonic steroids,[18] pharmaceutical compounds, such as fluticasone propionate and cefuroxime axetil,[19] plasma extracts of drug candidates[20] and DNA adducts.[21] Taylor and Teale[22] successfully used gradient CEC interfaced to ESI-MS for the analysis of benzodiazepines, corticosteroids and thiazide diuretics. Early reports[15,16,19] had commented on the problem of band-broadening in the un-packed capillary used for transfer into the mass spectrometer. Schmeer *et al.*[17] used capillaries that were fully packed, terminating inside the ESI source. Pressure was used in addition to EOF, so strictly this was a PEC technique, but comment was made that at higher field strengths, of 100 kV per m, flow velocity was almost independent of applied pressure. Lane *et al.*[23] devised an integrated system in which sample injection and separation were within the ESI probe interface, again with no un-packed section and allowing use of short columns with high field strengths for faster analyses. The use of fully packed, tapered capillaries for coupling to nanospray and low-flow ESI has been investigated[18] and Warriner *et al.*,[24] using a development of the integral probe developed by Lane *et al.*[23] have compared the sensitivity of nanospray and microspray ionization. Nanospray was found to be 5–10 times more sensitive than microspray, but band broadening was apparent, arising at the connector used to couple the CEC capillary with the narrower i.d. nanospray capillary. Open-tubular CEC has also been coupled with MS, using an on-line ion trap storage/reflectron time-of-flight (IT/reTOF), mass spectrometer, for the rapid analysis of peptides.[25] Six peptides were separated within 3 minutes using this system, relying on the fast scan rate of the mass spectrometer. The same group have also reviewed the coupling of HPLC as well as CE and CEC with IT/reTOF.[26] The particular advantage in using TOF analysers, as well as their sensitivity, resolution and mass range, is their fast scan duty cycle.

Matrix-assisted laser desorption ionization (MALDI)-MS has not yet been coupled with CEC, probably because it is not yet possible to connect it with a chromatographic technique in a dynamic way. Instead, fractions need to be collected for subsequent analysis by MS.

## CEC/MS Interfaces

As mentioned in the introduction, UV and fluorescence detection is performed in-line, actually on the un-packed section of capillary. However, detection using MS must be carried out upstream at the end of the capillary. In conventional CEC, using detection by light, the outlet of the capillary is immersed in a buffer vial, but to couple with MS the capillary outlet must be taken into an interface prior to the ionization source of the mass spectrometer. The interface must also provide electrical contact at the capillary outlet, and this is often at high voltage rather than at ground potential as with conventional CEC. Ionization sources in sector mass spectrometers have several kV accelerating potential and ESI spray capillaries usually operate at approximately 3–5 kV, although there are some instruments where the capillary is at ground potential. Consequently the voltage drop across the CEC capillary is the difference between the CEC separation voltage and the MS voltage.

The essential requirement of the interface is maintenance of the chromatographic separation efficiency through into the mass spectrometer. Several types of interface have been developed and these are described in detail below, with discussion of their relative merits. Coupling of CEC with MS uses essentially the same techniques as CE coupling, which has been performed for a longer period and by more researchers. The reader is encouraged to consult the literature on CE/MS, much of which has relevance to CEC/MS, as well as the literature on CEC/MS itself.

Since CEC is a liquid-phase separation technique and mass spectrometry separates ions in the gas phase, an ionization source is required in which the analyte is converted into the gas phase with simultaneous or subsequent ionization. The flow rate of the eluent in CEC, approximately 100 nl min$^{-1}$ is too low for conventional ion sources (nanospray is discussed later) and a 'make-up' flow of liquid must be used, to bring the total flow up to the 'normal' level. In the case of ESI, this would typically be 3–20 $\mu$l per minute; however, it is possible to consider using wider bore CEC capillaries, approximately 100–200 $\mu$m i.d., with a high enough flow rate for direct linking. The most common ionization technique now used for liquid chromatographic coupling, including CE and CEC, is electrospray and this is likely to remain the case for the foreseeable future, since it is inherently the most suitable and robust. The addition of organic solvent in the make-up liquid reduces the surface tension of more aqueous buffers to aid formation of charged droplets. This is more applicable to CE than to CEC, since the latter generally uses a high content of organic solvent.

There are several types of interfaces for coupling the CEC capillary with the ionization source and these are now discussed in more detail.

## Direct Coupling

The first successful coupling of CE with MS was described by Smith's group in 1987,[27] where a stainless steel capillary surrounded the CE capillary coaxially at the cathodic outlet end and acted as the electrospray needle and cathode. Capillaries with 100 $\mu$m i.d. were used to provide a sufficiently high flow rate for the electrospray process, but a relatively large dead volume at the outlet reduced efficiency.

## Coaxial Coupling

Following the work on direct coupling, Smith's group developed an interface using coaxial 'make-up', where a liquid sheath flow provided electrical contact as well as increasing the flow for ESI.[28–30] This design also had the advantage that the CE capillary could be extended to the tip of the probe interface, thus minimizing dead volume and maintaining chromatographic efficiency. As a result, this interface has become the most widely used by most groups, providing good sensitivity, maintenance of chromatographic efficiency and robustness in use. Smaller i.d. capillaries are used with this system, typically 50–75 $\mu$m, or

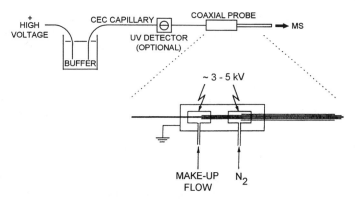

**Figure 6.4** *Schematic diagram of CEC/MS system with detail of coaxial electrospray probe*

even smaller. Figure 6.4 shows a schematic diagram of a CEC/MS system with a coaxial probe interface for a 'Micromass' ESI mass spectrometer, which may also be used for CE/MS. The interface essentially comprises two concentric stainless-steel capillaries, which surround the inner CEC capillary, and two stainless-steel T pieces. The sheath flow liquid is carried in the stainless-steel capillary which is concentric with the separation capillary within and the liquids only meet at the outlet point. The first T piece allows incorporation of 'make-up' flow, whilst the second T piece is used to supply nitrogen into the outer stainless steel capillary to nebulize the liquid. High voltage, approximately 3–5 kV, is applied to both T pieces, and thence to the stainless-steel capillaries for the electrospray process and providing electrical contact for the CEC column. Detection by UV may be carried out on-line, requiring an un-packed section of capillary, but the advantage of using a completely packed capillary, with no UV detection, is discussed later.

Tetler *et al.*[31] have studied the influence of the relative inner and outer diameters of capillaries in a coaxial CE/MS system, with applicability to CEC/MS and suggested an optimum arrangement of separation, make-up and nebulizing gas capillaries in terms of sensitivity and stability. However, this relies on the use of capillaries that are not readily available commercially, but does demonstrate that consideration must be given to these factors to achieve best operation.

Optimal ESI performance has been shown in CE/MS when CE and ESI electric currents are similar,[32] while in the case of CEC/MS, currents are even more equally matched.

Many of the problems in coupling CE using open capillaries with MS are obviated by the use of packed capillaries. For example, where the composition of the sheath make-up liquid is different from the separation buffer and contains different counterions, then back migration of these counterions into the separation capillary can affect performance, such as loss of resolution or change in elution order of analytes.[33] This is not so much of a problem in

CEC, where flow into the CEC capillary from the make-up flow is restricted by the presence of the packing material. Also, when coupled to high-vacuum sources such as FAB, vacuum-induced flow is a problem with open CE capillaries, but again this does not present a problem with packed CEC capillaries. A common problem with both CE and CEC, involving liquid sheath interfaces coupled with ESI, is dilution of analytes and competition for charge by ionizable species in the sheath solvent,[34] both effects reducing analyte sensitivity.

### Liquid-junction Interface

The liquid-junction interface was first described by Minard for CE[35] and involved immersion of the end of the separation capillary in buffer, which was at ground potential, and close alignment to the end of an MS transfer line capillary, with a gap of approximately 25 μm. Matrix was also added to the buffer for coupling to a CF-FAB source and the vacuum from the mass spectrometer pulled the CE eluent, FAB matrix and make-up liquid at a flow rate of 5 μl min$^{-1}$. Henion and co-workers[36] described a liquid-junction interface for use with ESI, involving use of a stainless-steel T piece, into which the CE capillary and transfer capillary were aligned opposite each other, with a gap of approximately 10–25 μm for incorporation of make-up flow from the third inlet of the T piece. Electrical connection was also made at the T piece for the CE system and to provide ESI voltage. Figure 6.5 shows a schematic diagram of a liquid-junction interface. The main drawback of this interface is the possible loss of chromatographic efficiency because of excessive dispersion at the gap between the capillaries. It is very difficult to align the capillaries correctly and reproducibly to minimize dispersion and therefore most workers prefer the coaxial interface.

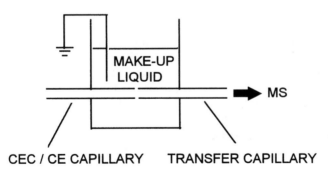

**MAKE-UP LIQUID**

**MS**

**CEC / CE CAPILLARY**    **TRANSFER CAPILLARY**

**Figure 6.5** *Schematic diagram of liquid-junction interface*

### Combined Coaxial/Liquid-junction Interface

Caprioli *et al.*[37] developed a combined coaxial/liquid-junction interface which incorporated aspects of the liquid-junction and coaxial interfaces and was used

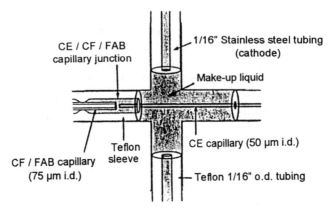

**Figure 6.6**  *Schematic diagram of combined coaxial/liquid-junction interface*
(Adapted from ref. 37)

for CE/CF-FAB. Figure 6.6 shows a schematic diagram of this interface, which essentially is an adjustable coaxial system within a 4-way coupling at which make-up flow and FAB matrix is added. The CE capillary may be placed at varying distances, as well as inside the CF-FAB transfer line, allowing easier optimization of the system than a liquid-junction alone. The interface was not widely adopted, probably as a result in the growth of ESI and use of coaxial systems.

### Nanospray and Microspray

As mentioned in the section on the coaxial interface, the use of a sheath flow probably results in a loss of sensitivity from the ideal and the removal of sheath flow, which is highly desirable, is possible by use of nanospray and microspray ionization, where the more recent developments in ESI have occurred. The flow rate in CEC is typically 100 nl min$^{-1}$, depending on capillary i.d., voltage and buffer, being ideally suited to nanospray which operates at these flow rates and shows enhanced sensitivity over conventional ESI.[38]

Chowdhury and Chait in 1991,[39] demonstrated that a needle with a fine taper would allow aqueous solutions to be analysed by ESI. Several groups then pursued the development of low-flow interfaces. Gale and Smith[40] used capillaries of 5–20 μm i.d. etched with hydrofluoric acid and coated with silver for electrical contact, initially for pumped flow and then for CE/ESI-MS coupling. Caprioli *et al.*[41] used the term 'micro-electrospray' to distinguish this miniaturized technique from conventional ESI, while the term 'nanospray' was used by Wilm and Mann,[38] relating to the low nanolitre per minute flow rate. The nanospray technique was developed to take advantage of long acquisition times from limited sample amounts, together with enhanced sensitivity, and unlike the other systems did not rely on a driven flow. The electrospray process itself produced flow from the tapered capillary into which the sample was

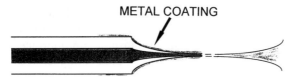

**Figure 6.7** *Schematic diagram of tapered CEC capillary for nanospray*

deposited and the efficiency was estimated to be two orders of magnitude higher than conventional ESI. There are two essential requirements for the nanospray process, the first being a small sprayer orifice and this is usually achieved, especially for single sample analysis not involving any separation, by the use of finely drawn borosilicate glass capillaries with tapered outlets of 1–3 $\mu$m. Figure 6.7 depicts a tapered CEC capillary for nanospray. The other requirement is that electrical contact must be made with the capillary tip and this is usually achieved by gold coating, which leads to the main problem with the technique, namely that the coating is not very robust. The sensitivity gain produced relates to the production of smaller droplets, approximately 200 nm diameter or less, compared to 1–2 $\mu$m for conventional ESI, from a smaller diameter capillary outlet, leading to improved ionization efficiency from the higher surface-to-volume ratio. A greater proportion of ions is likely to enter the mass analyser with nanospray, since the capillary is placed much nearer to the analyser entrance.

CEC coupled with nanospray and microspray has been investigated. We have packed 3 $\mu$m particles into 100 $\mu$m i.d. capillaries with an outlet taper approximately 10 $\mu$m i.d., also obviating the need for a retaining frit.[18] (Capillaries with smaller i.d. tapers would be much more prone to plugging and make packing difficult because of restricted flow rate.) Electrical contact was made by coating the tapered tip of the capillary with silver. Comparison of MS data from these capillaries with similar capillaries having a conventional blunt-ended tip and using coaxial make-up flow showed narrower showed narrower chromatographic peaks in the former. This was attributed to the absence of a dilution effect from make-up flow with the sheathless system. The lack of make-up flow allows no opportunity to alter the composition of the CEC eluent to foster the electrospray process, and this may cause problems (see practical aspects section). However, Severs *et al.*[42,43] have described a novel means of acidifying CE analytes at the CE/ESI-MS interface, by use of microdialysis tubing. Electrical connection was also made at the microdialysis junction, to enable sheathless, low-flow ESI. This technique would be equally suitable for CEC.

## CEC/MS Interfaces – Conclusions

The choice of the most suitable interface to couple CEC with mass spectrometers depends on the mode of ionization being used; as already stated it seems that ESI will continue to be the method of choice for the foreseeable future.

Suter and Caprioli[44] compared coaxial and liquid-junction interfaces for CE coupled with CF-FAB and found that the coaxial system attained the better electrophoretic performance, but the liquid-junction interface showed equal or better overall performance. This was attributed to longer analysis times in the latter because of vacuum-induced flow in the coaxial system, which is not a problem in CEC. Vacuum-induced flow with high-vacuum systems limits the i.d. of open CE capillaries typically to 10 $\mu$m, when using a coaxial system. Generally it was found that the liquid-junction interface was easier to handle and set up, although dead volume was a significant problem.

A comparison of coaxial and liquid-junction interfaces for CE coupled with ESI was made by Pleasance *et al.*,[45] who showed that the coaxial system provided the best performance and that the liquid-junction interface showed significant band broadening.

The overall conclusions are that for CEC/ESI coupling, a coaxial system will provide the best maintenance of separation efficiency and be reasonably straightforward to set up. However, the situation for CF-FAB is not so clear cut, where the main problem in our experience is the length of capillary required to pass through the FAB probe and to provide sufficient flexibility to enable manipulation through vacuum locks. A longer separation capillary would reduce the field strength across the column and result in excessively long retention times. A liquid-junction interface may have advantages in this case, and the combined coaxial/liquid-junction may be most suitable. The ideal system would appear to be the use of nanospray ionization, requiring no make-up flow, but this does have the disadvantage that it adds complexity, is not easy to set up and is not robust.

# 4   Practical Aspects

## General

A problem with CEC is bubble formation in the un-packed section of capillary, probably arising from flow velocity differences either side of the outlet frit. This need not be a problem when CEC is coupled to MS, since it is possible and desirable to dispense with the open section, the column terminating at the ionization point of the mass spectrometer. There is then no opportunity for post-column band broadening, although of course on-line UV detection is not then possible.

The narrow chromatographic peaks achievable by CEC must be scanned by the mass spectrometer and several scans (about 7), are required to characterize adequately a chromatographic peak. Scanning over a large $m/z$ range can be problematical because of relatively long MS cycle times, especially for sector instruments. Most reports to date, have used quadrupole instruments, although TOF has been used, with its significant advantages as noted earlier. Selected ion recording (SIR) and, with MS/MS instruments, multiple reaction monitoring (MRM) provide greater sensitivity (approximately 100 times) than scanning acquisitions. Figure 6.8 shows an example of a total ion chromatogram, with

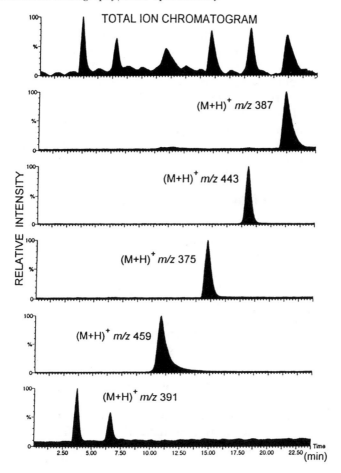

**Figure 6.8** *Total ion and extracted ion chromatograms* (Adapted from ref. 18)

extracted ion chromatograms at the appropriate $m/z$ for the separated compounds.

The major problem with CEC capillaries and MS interfacing is drying of the column outlet, since the capillary is not immersed in liquid. Consequently, steps must be taken to ensure that the end of the CEC column is kept wet when not flowing and we have found it useful when using a coaxial system to place an eppendorf tube or glass vial fitted with a rubber septum and filled with buffer over the tip of the CEC capillary/MS probe.

## CE/CEC Instruments

Commercially available CE/CEC instruments may be coupled with mass spectrometers and some manufacturers supply suitable interfaces and adapters.

The essential requirement is that the outlet end of the capillary is able to exit from the instrument without being excessively long, which may necessitate drilling holes in cabinets. Some instruments have current feedback circuits, which monitor the current across the capillary, requiring electrical connection with the outlet capillary. When coupling with MS this is not usually possible, and it may be necessary to disable the feedback circuit in order to stop the CE/ CEC instrument shutting off the separation voltage.

If UV detection is not being used, then the CE/CEC instrument is essentially acting only as an electrical power supply and an alternative is to just use a suitable high-voltage supply.

## Buffers and Sheath Liquids

Both the concentration and the flow rate of buffers in CEC is low, but non-volatile buffers are not recommended for long periods of use, since the performance of the mass spectrometer will deteriorate with time. However, non-volatile buffers may be used for short periods, especially when diluted by coaxial make-up flow and are much more compatible with orthogonal ESI sources. A particular problem with the use of nanospray is the lack of opportunity to modify or dilute the CEC eluent, and this is demonstrated in the two spectra in Figure 6.9. The spectrum in Figure 6.9(a) was from CEC/ESI-MS of a compound of molecular mass 386, using a sodium salt buffer, but using a coaxial make-up flow of methanol/0.1% formic acid in water, 1:1 v/v. The buffer has been diluted by the make-up flow, allowing the protonated molecule ion to predominate. However, Figure 6.9(b) was from CEC/nanospray-MS of the same sample and shows a predominant $[M + Na]^+$ ion. The problem with cationization by alkali metals is that there is competition for ion current between all the analyte adduct ions, including the protonated molecule ion and this reduces overall sensitivity. Conversely, it may actually be useful to add more of a competing cation, to improve sensitivity by fostering ionization of a particular species.

Ammonium acetate is particularly recommended as a suitable buffer for CEC/MS, together with methanol/water for the make-up sheath liquid (methanol has been found to be more compatible with the polyimide capillary coating than acetonitrile). Both buffer and make-up liquid should be filtered and degassed by ultrasonication before use. Sheath liquid may usefully be used to modify the solvent composition for optimum MS ionization, for example incorporation of a volatile organic acid such as formic or acetic acid at approximately 0.1% (v/v) to aid protonation for acquisitions in positive mode. Conversely, volatile bases may be used to promote deprotonation in negative mode, while ammonium acetate is suitable for either.

Figure 6.10 is a graph of MS chromatographic peak width and sensitivity against flow rate of coaxial make-up sheath liquid, using a 'Micromass' 'Quattro' mass spectrometer with ESI and coaxial probe, showing that there is an optimum flow rate where peak width is minimal and sensitivity is maximal. This would depend on the particular instrumental set-up, but in this case would

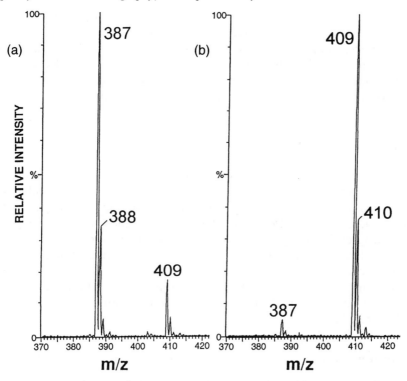

**Figure 6.9** (a) *Spectrum of compound molecular mass 386 from CEC/ESIMS with coaxial make-up.* (b) *Spectrum of same compound using nanospray ionization* (Adapted from ref. 18)

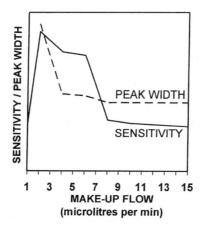

**Figure 6.10** *Graph of MS chromatographic peak width and sensitivity against coaxial make-up flow rate*

appear to be approximately 4–6 $\mu$l min$^{-1}$. Presumably higher flows cause too much dilution of analyte and reduce sensitivity. A flow rate greater than approximately 2–3 $\mu$l min$^{-1}$ is required to sweep the analyte clear of the probe tip and minimize peak width. In our experience, the tip of the CEC capillary should extend beyond the sheath liquid capillary by approximately 0.5 mm for best performance, but again this would need to be determined for the particular instrument set-up.

## Capillaries

Capillaries with an outer diameter of 375 $\mu$m are commonly used in CEC, for robustness and ease of use, but this is larger than the stainless-steel capillaries of conventional ESI probes. Therefore the mass spectrometer company's CE probe should be used or else the standard probe supplied with the instrument may be adapted by using larger i.d. stainless-steel capillaries to function as sheath capillaries for make-up flow and nebulization. An additional T piece could also be required.

The polyimide coating should be removed from approximately the last 4 mm of the outlet capillary, to facilitate contact with the sheath liquid and produce good ESI spraying.

## Sample Injection

Most researchers have used electrokinetic application of sample in CEC/MS, with the proviso that discrimination of charged analytes will occur. The ESI voltage is best turned off during sample application, as when the CEC voltage is switched off the ESI voltage would cause reversal of the EOF and may wash analytes back into the buffer. A further problem is the likelihood of drying the cathodic end of the CEC capillary.

## MS/MS Capability

MS/MS acquisitions may be performed on suitable instruments and a useful feature of electrically driven separation systems is that the voltage may be switched off for a few seconds to stop (actually reduce), the flow to enable further MS scans of an eluting analyte to be performed. The comment regarding reversal of EOF mentioned above does need to be borne in mind though, as the flow is only reduced rather than actually stopped for a few seconds as the separation voltage falls.

## 5  Future Prospects

There are several exciting developments in prospect for the future of CEC/MS and these include the coupling of microchip separations. The combination of CEC and TOF analysers seems likely to increase and will be appropriate for the increasing demand for high-throughput MS for analysis of combinatorial libraries and drug screening. The availability of commercial interfaces specifically for CEC/MS would be advantageous, as would the incorporation of photodiode array detection, and the use of fibre-optic technology seems likely in this area. This would further extend the analytical capabilities of the technique and may allow analysis of previously intractable mixtures, especially in the biological field where new compounds and biochemical pathways await discovery.

# References

1. M. Barber, R.S. Bordoli, R.D. Sedgwick, and A.N. Tyler, *J. Chem. Soc., Chem. Commun.*, 1981, 325.
2. M. Barber, R.S. Bordoli, G.V. Garner, D.B. Gordon, R.D. Sedgwick, L.W. Tetler, and A.N. Tyler, *Biochem. J.*, 1981, **197**, 401.
3. R.M. Caprioli, T. Fan, and J.S. Cottrell, *Anal. Chem.*, 1986, **58**, 2949.
4. Y. Ito, T. Takeuchi, D. Ishi, and M. Goto, *J. Chromatogr.*, 1985, **346**, 161.
5. M. Dole, L.L. Mach, R.L. Hines, R.C. Mobley, L.D. Ferguson, and M.B. Alice, *J. Chem. Phys.*, 1968, **49**, 2240.
6. C.M. Whitehouse, R.N. Dreyer, M. Yamashita, and J.B. Fenn, *Anal. Chem.*, 1985, **57**, 675.
7. A.P. Bruins, T.R. Covey, and J.D. Henion, *Anal. Chem.*, 1987, **59**, 2642.
8. G. Guiochon and H. Colin, in 'Proceedings of the 2nd International Congress on Techniques in Environmental Chemistry', ed. J. Albaiges, Barcelona, Spain, Nov. 1981, Pergamon Press, Oxford, 1982, 169.
9. T. Covey, in 'Biological and Biotechnological Applications of Electrospray Ionisation Mass Spectrometry', ed. A.P. Snyder, ACS Symposium Series 619, American Chemical Society, Washington, DC, 1996, 21.
10. A.L. Burlingame, R.K. Boyd, and S.J. Gaskell, *Anal. Chem.*, 1998, **70**, 647R.
11. E.R. Verheij, U.R. Tjaden, W.M.A. Niessen, and J. van der Greef, *J. Chromatogr.*, 1991, **554**, 339.
12. M. Hugener, A.P. Tinke, W.M.A. Niessen, U.R. Tjaden, and J. van der Greef, *J. Chromatogr.*, 1993, **647**, 375.
13. S.E.G. Dekkers, U.R. Tjaden, and J. van der Greef, *J. Chromatogr. A*, 1995, **712**, 201.
14. T. Tsuda, *Anal. Chem.*, 1987, **59**, 521.
15. D.B. Gordon, G.A. Lord, and D.S. Jones, *Rapid Commun. Mass Spectrom.*, 1994, **8**, 544.
16. G.A. Lord, D.B. Gordon, L.W. Tetler, and C.M. Carr, *J. Chromatogr. A*, 1995, **700**, 27.
17. K. Schmeer, B. Behnke, and E. Bayer, *Anal. Chem.*, 1995, **67**, 3656.
18. G.A. Lord, D.B. Gordon, P. Myers, and B.W. King, *J. Chromatogr. A*, 1997, **768**, 9.
19. S.J. Lane, R. Boughtflower, C. Paterson, and T. Underwood, *Rapid Commun. Mass Spectrom.*, 1995, **9**, 1283.
20. C.J. Paterson, R.J. Boughtflower, D. Higton, and E. Palmer, *Chromatographia*, 1997, **46**, 599.
21. J. Ding and P. Vouros, *Anal. Chem.*, 1997, **69**, 379.
22. M.R. Taylor and P. Teale, *J. Chromatogr. A*, 1997, **768**, 89.
23. S.J. Lane, R. Boughtflower, C. Paterson, and M. Morris, *Rapid Commun. Mass Spectrom.*, 1996, **10**, 733.
24. R.N. Warriner, A.S. Craze, D.E. Games, and S.J. Lane, *Rapid Commun. Mass Spectrom.*, 1998, **12**, 1143.
25. J.-T. Wu, P. Huang, M.X. Li, M.G. Qian, and D.M. Lubman, *Anal. Chem.*, 1997, **69**, 320.

26. J.-T. Wu, M.G. Qian, M.X. Li, K.F. Zheng, P.Q. Huang, and D.M. Lubman, *J. Chromatogr. A*, 1998, **794**, 377.
27. J.A. Olivares, N.T. Nguyen, C.R. Yonker, and R.D. Smith, *Anal. Chem.*, 1987, **59**, 1230.
28. R.D. Smith, C.J. Barinaga, and H.R. Usdeth, *Anal. Chem.*, 1988, **60**, 1948.
29. J.A. Loo, H.R. Usdeth, and R.D. Smith, *Anal. Biochem.*, 1989, **179**, 404.
30. R.D. Smith, H.R. Usdeth, C.J. Barinaga, and C.G. Edmonds, *J. Chromatogr.*, 1991, **559**, 197.
31. L.W. Tetler, P.A. Cooper, and B. Powell, *J. Chromatogr. A*, 1995, **700**, 21.
32. A.J. Tomlinson, L.M. Benson, and S. Naylor, *J. Cap. Electroph.*, 1994, **1**, 127.
33. F. Foret, T.J. Thompson, P. Vouros, B.L. Karger, P. Gebauer, and P. Bocek, *Anal. Chem.*, 1994, **66**, 4450.
34. J.H. Wahl, D.R. Goodlett, H.R. Usdeth, and R.D. Smith, *Electrophoresis*, 1993, **14**, 448.
35. R.D. Minard, D. Chin-Fatt, P. Curry, and A.G. Ewing, in 'Proceedings of the 36th Annual Conference on Mass Spectrometry and Allied Topics', San Francisco, CA, 1988, 950.
36. E.D. Lee, W.M. Muck, J.D. Henion, and T.R. Covey, *Biomed. Environ. Mass Spectrom.*, 1989, **18**, 844.
37. R.M. Caprioli, W.T. Moore, M. Martin, B.B. DaGue, K. Wilson, and S. Moring, *J. Chromatogr.*, 1989, **480**, 247.
38. M. Wilm and M. Mann, *Anal. Chem.*, 1996, **68**, 1.
39. S.K. Chowdhury and B.T. Chait, *Anal. Chem.*, 1991, **63**, 1660.
40. D.C. Gale and R.D. Smith, *Rapid Commun. Mass Spectrom.*, 1993, **7**, 1017.
41. M.R. Emmett and R.M. Caprioli, *J. Am. Soc. Mass Spectrom.*, 1994, **5**, 605.
42. J.C. Severs, A.C. Harms, and R.D. Smith, *Rapid Commun. Mass Spectrom.*, 1996, **10**, 1175.
43. J.C. Severs and R.D. Smith, *Anal. Chem.*, 1997, **69**, 2154.
44. M.J-F. Suter and R.M. Caprioli, *J. Am. Soc. Mass Spectrom.*, 1992, **3**, 198.
45. S. Pleasance, P. Thibault, and J. Kelly, *J. Chromatogr.*, 1992, **591**, 325.

CHAPTER 7

# Pharmaceutical Applications of Capillary Electrochromatography

MELVIN R. EUERBY AND NICOLA C. GILLOTT

## Preface

The new and exciting hybrid technique of capillary electrochromatography (CEC) which seeks to exploit the combined advantages of both capillary electrophoresis (high efficiencies) and high-performance liquid chromatography (mobile and stationary phase selectivity) is reviewed and discussed in relation to its current status and future prospects within the pharmaceutical industry. Even though the technique is still in its infancy with regards to instrumentation and packed capillary technology, there have been numerous, and increasingly frequent, reports of excellent separations, with the pharmaceutical industry being a major initiator in both fundamental research and the development of the technique.

## 1  Introduction

Since the re-emergence of CEC in the early 1990s, workers have mainly concentrated on establishing reliable column packing technologies[1-6] and investigating the theory and mechanisms underpinning CEC.[7-11] Only in the last few years has there been a steady flow of presentations and publications describing applications of CEC in the area of pharmaceutical analysis.

The pharmaceutical industry has been one of the driving forces for the development of CEC as the technique offers the potential for an orthogonal separation mechanism to the ubiquitous HPLC. With its highly efficient and rapid separations, and hence increased peak capacity, CEC is an attractive adjunct to conventional chromatographic analysis. Owing to the scarcity of specifically designed stationary phases, most of the CEC applications have been in the analysis of neutral and ion-suppressed species where the analyst is utilizing the increased efficiency and speed of analysis associated with CEC and not the orthogonal nature of the technique.

After a brief introduction to the current scope, applicability with problems associated with analysing various chemical functionalities, this chapter will then discuss the analysis of pharmaceuticals.

# 2   Scope and Applicability of CEC

## Neutrals

To date, most of the reported applications of CEC have involved the analysis of neutral species of widely differing structures. The pH of the mobile phase has been typically in excess of 7 in order to promote a high EOF due to increased silanol ionization. Traditional HPLC stationary phases, such as Hypersil ODS and Spherisorb ODS1, have generally been employed as they are unendcapped and possess a relatively large number of acidic silanol groups. In marked contrast, many pharmaceutical compounds possess an ionizable functionality and different approaches must be used for their analyses as discussed below.

## Acids

Acidic analytes that are analysed in their ionized form tend to migrate towards the anode, *i.e.* against the EOF, and either are not loaded onto the column during electrokinetic injection, or are not swept towards the detector and hence are not detected. In order to chromatograph acids successfully by CEC a mobile phase pH has to be employed which will chromatograph the acids in their ion-suppressed mode (*i.e.* two pH units below their p$K_a$). As a consequence of using acidic mobile phases, the acidic analyte behaves as a neutral component and the EOF, and hence linear velocities, are reduced. For example, typical linear velocities of 0.75 mm s$^{-1}$ have been reported at pH 2.3 compared to 1.5 mm s$^{-1}$ at pH 7.8. Despite the reduced EOF, successful and rapid analysis of acidic pharmaceuticals is still possible.

Recently, the rapid analysis of acids in their ion-suppressed mode has been reported[12] using mixed mode phases which possess a C-alkyl and sulfonic acid group bound to the same silica particle. The presence of the sulfonic acid group, which is ionized at all workable pHs, generates a good EOF over a wide pH range (*e.g.* linear velocity of 2.5 mm s$^{-1}$ at pH 2.5) thus enabling extremely rapid analysis to be performed without sacrificing the partitioning capacity of the phase.

## Bases

The perception that strong basic analytes, especially pharmaceutical drugs cannot be successfully analysed on reversed-phase materials currently used for CEC has significantly slowed the progress of the technique.

The analysis of the basic analytes using CEC is thought to be problematic because, in order to generate a substantial EOF, an acidic silica is required. These silanol groups, which are vital for the generation of the EOF, are detrimental to the CEC analysis of basic analytes, as severe peak tailing occurs owing to the mixed mode separation mechanism. The reversed-phase materials currently used for CEC were originally developed in the 1970s as HPLC phases; they too produced excessive peak tailing when used in HPLC. In order to minimize these interactions with the basic analytes, and hence improve the peak shape, small basic compounds such as triethylamine (TEA) were incorporated into the mobile phase, acting in a competitive manner so as to restrict the access of the basic analytes to the silanol groups on the surface of the silica. This approach has recently been shown to work successfully for the CEC analysis of pharmaceutical bases.[13–15] Figure 7.1 illustrates the marked improvement in peak shape on the additions of TEA to the mobile phase for a potential drug substance with a p$K_a$ of 8.3.

An alternative approach has been to analyse the basic analyte in its ion-suppressed mode, although this approach is limited by the p$K_a$ of the base and by the instability of the stationary phase in the alkaline conditions required. However, reasonable success has been reported for the analysis of a basic drug candidate possessing a p$K_a$ of 8 at a pH of 9.3 on a CEC Hypersil $C_{18}$ phase.[16]

Bases, not unexpectedly, have been found to exhibit poor peak shape at pHs on the mixed mode phases as observed with traditional ODS materials.[17]

Smith and Evans[1] reported an apparently elegant solution to the analysis of basic drugs by using a strong cation exchange (SCX) stationary phase. An 'on-column focusing' phenomenon of the bases produced efficiencies of up to

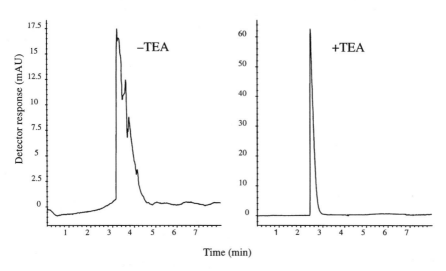

**Figure 7.1** *Electrochromatograms (terminology used as both partitioning and electro-mobility parameters in operation) for a basic Astra Charnwood research compound with and without TEA. Mobile phase, MeCN:0.05 mol dm$^{-3}$ NaH$_2$PO$_4$, pH 2.3:H$_2$O (6:2:2 v/v/v) with and without 0.1% TEA v/v. Sample concentration was 100 μg ml$^{-1}$. Electrochromatography was performed on a 25 cm, 100 μm i.d., 3 μm Hypersil phenyl packed capillary (Reprinted with permission from ref. 15)*

$8 \times 10^6$ plates m$^{-1}$ for the analysis of tricyclic antidepressants, whereas concomitant analysis of neutral species only produced efficiencies typically seen in reversed-phase CEC. These staggering efficiencies have also been obtained by other workers for a range of structurally diverse bases. However, all workers have experienced severe non-reproducibility of the phase, in that severe tailing and fronting have been unexpectedly observed in the middle of successful runs;[18-20] this phenomenon still remains to be explained.

## Simultaneous Analysis of Acids, Bases and Neutrals

For the simultaneous analysis of acids, bases and neutrals in a single CEC analysis, the use of TEA phosphate and triethanolamine phosphate at pH 2.5 in the mobile phase has been employed.[15] This approach allows the acids to be chromatographed in their ion-suppressed mode and, hence, run after the EOF with the neutral species and be separated by their differential partitioning between stationary and mobile phases. In contrast, the basic analytes elute before the EOF owing too their charged nature. Figure 7.2 shows the simultaneous analysis of an acidic (benzoic acid), a neutral (caffeine) and a basic (benzylamine) compound. The analysis was achieved within 8 minutes with acceptable chromatography obtained for all three components.

This approach makes it possible to analyse acids and bases, or bases and neutrals, or acids, bases and neutrals in a single CEC run. Adoption of this technique should extend the applicability of CEC to the analysis of more complex mixtures of differing chemical diversity. In contrast, acid and neutral

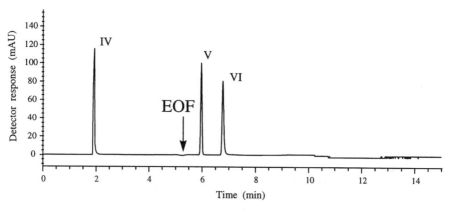

**Figure 7.2** *Electrochromatograms of benzylamine (IV), caffeine (V) and benzoic acid (VI). Efficiency values of 4642, 76331 and 63299 plates per column and symmetry values of 2.54, 0.90 and 0.71 were obtained respectively for (IV), (V) and (VI). Electrochromatography was performed at an applied voltage of 30 kV on a 25 cm, 100 μm i.d., 3 μm Hypersil phenyl packed capillary, mobile phase, MeCN:0.10 mol dm$^{-3}$ triethanolamine phosphate, pH 2.5:H$_2$O (6:2:2 v/v/v). Sample concentration was 100 μg ml$^{-1}$ of each component* (Reprinted with permission from ref. 15)

components can be simultaneously analysed by using a mobile phase of suitable pH to maintain the acid in the ion suppressed mode, and a traditional $C_{18}$ stationary phase material or one of the newer mixed mode phases ($C_{18}$/SCX or $C_6$/SCX).

# 3 Pharmaceutical Application of CEC

## Neutral Pharmaceuticals

### Overview and Scope

Given the current limitations of the technique, CEC has found a stronghold in the analysis of neutral or ion-suppressed drug substances and intermediates of wide structural diversity. Cephalosporin antibiotics,[21,22] prostaglandins,[21] diuretics,[1,18,23] steroids,[1,18,22,24-26] macrocyclic lactones,[24] C- and N-protected peptides,[18,27] nucleosides and purine bases,[18] phthalates,[7] and parabens[8] have been successfully chromatographed by CEC using traditional $C_{18}$ stationary phases, usually with a mobile phase containing a high proportion of acetonitrile and a pH in the range of 2.3 to 9.3.

CEC is particularly useful in the early stages of drug discovery, where rapid method development is essential and the demands of validation are not so tight. In nearly all cases for neutral and acidic compounds run in their ion-suppressed mode separations were more efficient and rapid, and the method development time was dramatically reduced. Two C- and N-protected tetrapeptides that differ only in the methylation of one amide function were found to be separated in a run time of less than 4 minutes, using the standard CEC test chromatographic conditions that are employed to check the performance of our capillaries. In comparison, the HPLC method previously used involved a 30-min gradient. Therefore, the CEC method was the obvious choice as the quantitative results from both techniques were comparable.[18]

Method development in CEC is easily automated on commercially available CEC instrumentation, and the chromatographic theory central to the computer optimization in HPLC is valid in CEC for analytes run in their ion-suppressed mode.[18] This enables method development to progress at a rapid pace. Riley *et al.*[28] have extended this work to show the usefulness of a modified central composite design to optimize the CEC separation of the antibacterial 3-[4-methylsulfinyl)phenyl]-5*S*-acetamidomethyl-2-oxazolidinone from its three related S-oxidation products. The variables included in the investigation were applied potential, volume fraction of acetonitrile (MeCN) and buffer concentration. The end result was the development of a rugged CEC method for the separation of the antibacterial from its thioether, sulfone and sulfoxide diastereoisomer on a 3 $\mu$m CEC Hypersil ODS phase in 9 minutes.

Recent work performed by Miyawa *et al.*[29] for the separation of an anti-inflammatory and its related impurities, used CEC as a method development tool for the liquid chromatographic separation. As long as the potential differences[30] between the techniques are kept in mind it is possible to develop LC methods using CEC and *vice versa*.

## Steroids

Steroids seem to be particularly amenable to analysis by CEC, a typical example being that of the separation of the corticosteroid tipredane from its diastereo-isomer and four related substances by CEC using the standard CEC conditions with an unpressurized system.[18] As can be seen in Figure 7.3, baseline separation of tipredane from its diastereoisomer was achieved with no method development; in contrast, HPLC failed to produce baseline separation of the diastereoisomers, even after extensive stationary and mobile phase optimiza-tion. Stead *et al.*[31] also used a steroid, progesterone, to assess whether CEC can be used to separate compounds from a complex biological matrix. On-line concentration was used in order to offset the low concentration sensitivity of CEC by injecting from a non-eluting solution. The method was successfully transferred from an existing HPLC method and found to provide gains in both efficiency and speed of analysis.

Boughtflower's and Smith's groups have also shown that CEC is highly effective for the separation of synthesis-related impurities in the corticosteroid fluticasone.[21,24] Lord *et al.*[32] have highlighted the use of CEC in the separation of glycosolated steroids such as bufadienolide (bufalin, cinobufagin and cinobufatakin) and cardenolide (digoxigenin, gitoxigenin and digitoxigenin).

There have been numerous reports of the separation of endogenous steroids such as hydrocortisone, testosterone, 17-α-methyltestosterone, progesterone[33-37]

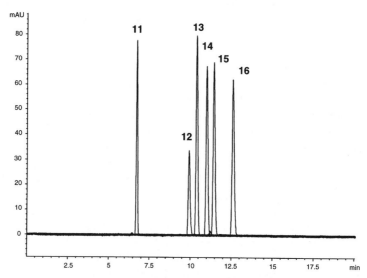

**Figure 7.3**  *Separation of tipredane (14) from its diastereoisomer (15) and structural analogues (11–13, 16). Electrochromatography was performed on an unpres-surised HP³ᴰ CE system using a 3 μm Spherisorb ODS 1 column (25 cm, 50 μm i.d., MeCN:0.05 mol dm⁻³ TRIS, pH 7.8 (8:2 v/v) at an applied voltage of 30 kV and capillary temperature of 15 °C*
(Reprinted with permission from ref. 18)

and many synthetic corticosteroids such as triamcinolone, hydrocortisone, prednisolone, cortisone, methylprednisolone, betametasone, dexamethasone, adrenosterone, fluocortolone and triamcinolone acetonide by CEC.[34,38] Most have reported higher efficiencies associated with these separations.

Ross *et al.*[34] further highlighted the usefulness of CEC in the separation of diastereoisomers in that they recently showed the excellent separation of the α and β isomers of 17-hydroxycholesterol using conditions similar to those described above.

## Acidic Pharmaceuticals

### *Diuretics*

Taylor and Euerby[18,38] have both reported good chromatography of thiazide diuretics using CEC. In order to achieve the CEC analysis of the acidic thiazides, a mobile phase pH of 2.5 must be employed in order to chromatograph the acids in their ion-suppressed mode.[18] As can be seen in Figure 7.4(a), six thiazide diuretics could be successfully analysed.

### *Barbiturates*

A systematic investigation of the CEC separation of six barbiturates was investigated on commercially available $C_{18}$, $C_8$ and phenyl bonded packing materials.[12,40] The effect on retention behaviour of mobile phase composition, buffer pH, buffer ionic strength and temperature were studied.

The separation of the barbiturates was optimized on the Hypersil phenyl phase in terms of pH, ionic strength and organic content of the mobile phase. A comparison of the barbiturate analysis using optimized conditions on three different packing materials is shown in Figure 7.5. Increasing the mobile phase pH resulted in a more rapid elution of the barbiturates as did increasing the percentage organic in the mobile phase. Decreasing buffer concentration from $0.05 \text{ mol dm}^{-3}$ to $0.00625 \text{ mol dm}^{-3}$ (pH 4.5) using the Hypersil phenyl stationary phase also resulted in a reduction in the retention time.

Increasing the temperature of the separation had the effect of reducing the retention time of all the analytes without changing the order of elution.

### *Non-steroidal Anti-inflammatory Drugs (NSAIDS)*

One of the problems associated with analysing acids in their ion-suppressed mode is the need for a low-pH mobile phase resulting in a significantly reduced EOF, which leads to longer analysis times. In order to circumvent this problem, several new phases have been developed which incorporate a SCX functionality, *e.g.* propylsulfonic acid moiety and a C-alkyl ligand on the same silica particle. The SCX functionality generates an EOF which is independent of the mobile phase pH used.

Preliminary results using the mixed mode phase containing a $C_{18}$ and

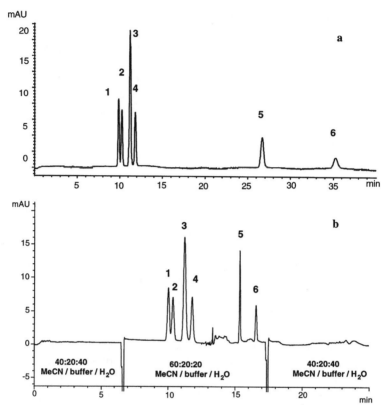

**Figure 7.4**  *CEC separation of the diuretics chlorothiazide (1), hydrochlorothiazide (2), chlorothalidone (3), hydroflumethiazide (4), bendroflumethiazide (5) and bumetanide (6). (a) Isocratic separation, CEC Hypersil $C_{18}$ column (25 cm, 50 μm i.d.), MeCN:0.05 mol dm$^{-3}$ NaH$_2$PO$_4$, pH 2.5: H$_2$O (6:2:2 v/v/v). (b) Step-gradient, column as for (a); 0–6.50 min, MeCN:0.05 mol dm$^{-3}$ NaH$_2$PO$_4$, pH 2.5:H$_2$O (4:2:4 v/v/v); 6.50–17.25 min, (6:2:2 v/v/v), 17.25– 25.00 min, (4:2:4 v/v/v)*
(Reprinted with permission from ref. 39)

propylsulfonic moiety ($C_{18}$/SCX) produced good partitioning and high linear velocities over a wide pH range. The linear velocity obtained was largely independent of the mobile phase pH (2.3–7.8) and was substantially higher than for the conventional $C_8$, $C_{18}$ and phenyl bonded phases.[12] The use of this experimental phase for the separation of a range of acidic NSAIDS is illustrated in Figure 7.6.

## Cannabinoids

Lurie *et al.*[41] have demonstrated the applicability of CEC to the analysis of a range of biologically active cannabinoids, which are acidic in nature and the

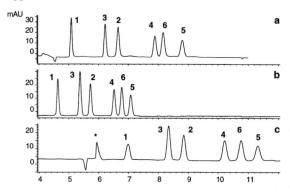

**Figure 7.5** *Electrochromatograms of the separation of the optimized separation of the barbiturates, barbital (1), butobarbital (3), phenobarbital (2), amobarbital (4), secobarbital (5) and hexabarbital (6) on three different packing materials. (a) 3 μm Hypersil $C_8$, mobile phase, MeCN:0.05 mol dm$^{-3}$ MES, pH 6.1:$H_2O$ (5:2:3 v/v/v), 15 °C; (b) 3 μm CEC Hypersil $C_{18}$, mobile phase, MeCN:0.05 mol dm$^{-3}$ MES, pH 6.6:$H_2O$ (4:2:4 v/v/v), 15 °C; (c) 3 μm Hypersil phenyl, mobile phase, MeOH:0.05 mol dm$^{-3}$ NaH$_2$PO$_4$, pH 4.5:$H_2O$ (5:2:3 v/v/v), 60 °C. Electrochromatography was performed at an applied voltage of 30 kV and 8 bar pressure with packed capillaries 25 cm 100 μm i.d.*
(Reprinted with permission from ref. 40)

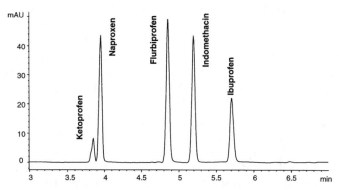

**Figure 7.6** *Isocratic CEC separation of NSAIDS on a CEC Hypersil mixed mode phase ($C_{18}$/SCX), 210 mm × 50 μm i.d., 3 μm packed capillary; MeCN:NaHPO$_4$ (50 mmol/l, pH 2.3): $H_2O$, (6:2:2 v/v/v); 210 nm, 15 °C, 30 kV, 5 kV/15 second injection, voltage ramp 0.5 min, 8 bar capillary pressurization, 1 s DAD rise time, analyte concentration 0.24 mg ml$^{-1}$, except ketoprofen 0.04 mg ml$^{-1}$ in MeCN:$H_2O$ (1:1 v/v)*
(Reprinted with permission from ref. 12)

major constituents of marijuana and hashish, both major drugs of abuse. The baseline separation of seven cannabinoids was achieved on a 100 $\mu$m i.d. CEC Hypersil $C_{18}$ capillary with a mobile phase consisting of 75:25–MeCN:0.025 mol dm$^{-3}$ NaH$_2$PO$_4$, pH 2.57 with an analysis time of just under 15 minutes.

Once again the low pH of the mobile phase was essential to allow the acidic cannabinoids to be chromatographed in an ion-suppression mode. The CEC analysis was found to resolve approximately 50% more peaks for a hashish sample than an HPLC analysis with a similar run time.

## Basic Pharmaceuticals

Owing to the severe problem of peak tailing associated with the analysis of bases using traditional reversed-phase silica materials there have been limited examples published. Taylor and Teale[38] reported the analysis of two benzodiazepines – diazepam and nitrazepam – using a standard Hypersil ODS phase and an ammonium acetate:MeCN mobile phase. Under isocratic conditions the resultant peaks were quite broad and exhibited peak tailing. However, under gradient conditions the peaks were found to be more Gaussian in appearance owing to the gradient effect on the tail of the peak.

Following the initial euphoria and excitement associated with the findings of Smith and Evans[1] regarding the success of analysing bases on Spherisorb SCX material, numerous workers have investigated the use of SCX materials for the analysis of a wide range of structurally diverse bases such as the alkaloids palmatine and jatrorrhuzine,[42] purines, pyrimidines, serotonine derivatives and peptides.[43]

Owing to the non-reproducible nature of the chromatographic mechanism, results have been varied. In order to further investigate this anomaly, Smith's group have synthesized a range of SCX phases with differing base silicas and differing linkers. Results from these new phases have been mixed, focusing and non-reproducibility still being exhibited.[19,44] The Waters Symmetry SCX experimental phase has been reported to be successful for the analysis of a range of pharmaceutical compounds, including the basic analyte propanolol and the neutral components *E*- and *Z*-isomers and tamoxifen, ibuprofen and carbovir with reasonable peak shape.[19]

Another approach investigated by Stobaugh *et al.*[45] has been to analyse bases on bare silica. Interestingly, the same type of focusing was observed with nortriptyline, amitriptyline and imipramine using a 1.5 $\mu$m Alltech silica and a 80:20 – MeCN:0.004 mol dm$^{-3}$ (NH$_4$)$_2$HPO$_4$, pH 8.0 mobile phase. However, the method still suffered from being non-reproducible. To date, the most successful method of analysing basic analytes by CEC involves the incorporation of competing bases into the mobile phase on traditional reversed-phase materials using an acidic mobile phase.[15] A wide range of pharmaceutical bases, including procainamide, nortriptyline and diphenhydramine, and a range of potential drug candidates with p$K_a$ values of 8 and above have been successfully analysed by CEC on commercially available packed capillaries with excellent peak shapes and efficiency values as seen in Figure 7.7. The bases, owing to their electromobility, elute before the EOF as can be seen in Figure 7.7. It is interesting to note that the elution/migration order of the basic analytes ran under these CEC conditions is different from that of CE and HPLC run under identical mobile phase conditions, see Table 7.1. This highlights the orthogonal nature of CEC when both partitioning and electromobility mechanisms act together.

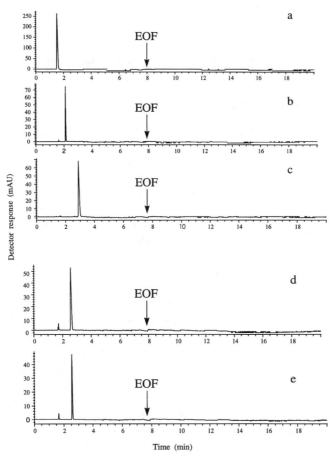

**Figure 7.7** *Electrochromatograms of nortriptyline* (a) *benzylamine* (b) *procainamide* (c) *diphenhydramine* (d) *and a basic Astra Charnwood research compound* (e). *Efficiency values of 2532, 10 469, 10 269, 8288 and 21 434 plates per column were obtained respectively for* (a), (b), (c), (d) *and* (e). *Conditions as stated in Figure 7.2 except: mobile phase, MeCN:0.15 mol dm$^{-3}$ TEA phosphate, pH 2.5:H$_2$O (6:2:2 v/v/v). Sample concentration was 10 µg ml$^{-1}$*
(Reprinted with permission from ref. 15)

**Table 7.1** *Elution order of bases when analysed using similar conditions by three chromatographic techniques*

| Compound | Elution order | | |
|----------|------|-------|-----|
|          | *CEC* | *HPLC* | *CE* |
| Benzylamine | 1 | 1 | 1 |
| Astra drug candidate 1 | 2 | 4 | 5 |
| Procainamide | 3 | 3 | 6 |
| Astra drug candidate 2 | 4 | 2 | 3 |
| Diphenhydramine | 5 | 5 | 2 |
| Nortriptyline | 6 | 6 | 4 |

# 4   Issues Associated with CEC in Pharmaceutical Analysis

## Analysis of Pharmaceutical Formulations

To date there has been limited work in this area mainly owing to the fear that repeated injections of pharmaceutical excipients may cause premature deterioration of these expensive packed capillaries. As the following reports demonstrate, pharmaceutical excipients do not necessarily cause run failure due to poisoning of the phases. However, careful consideration of the workup procedure must be made as it should also in HPLC. Demarest *et al.*[46] have reported the analysis of a neutral and very hydrophobic pharmaceutical in capsules containing the formulation excipients poly(vinylpyrrolidone), sodium dodecylsulfate and magnesium sulfate using a simple sample preparation methodology with a Hypersil $C_8$ packed capillary and a 60:40 – MeCN:TRIS mobile phase. The use of an internal standard in the extraction solvent yielded a 0.47% relative standard deviation (RSD) for the peak area ratio. The CEC and HPLC results were shown to be comparable. Adam *et al.*[47] have reported the development of a rapid and reliable method for the separation of hydrocortisone derivatives and preservatives from creams and ointments using a $C_{18}$/SCX mixed mode phase. Prior to analysis an unspecified workup procedure was employed. The use of non-aqueous CEC mobile phases such as MeCN, isopropanol, *n*-hexane and ammonium acetate combined with a CEC Hypersil 3 $\mu$m ODS material has been successfully applied to the analysis of testosterone esters in a peanut oil-based pharmaceutical formulation. This approach has also been applied to the separation of numerous pharmaceutical excipients such as triglycerides from over 30 different sources such as vegetable oils, foods, and soya lecithin extracts.[48]

## Analysis of Related Impurities in Drug Substances

The determination of related impurities in a drug substance has been traditionally performed by reversed-phase liquid chromatography where limits of quantitation below 0.05%, which are imposed on the pharmaceutical industry by regulatory agencies,[49] can be easily achieved in most instances.

A corticosteroid was used to evaluate the suitability of CEC for the determination of related impurities from bulk drug substance.[50] It was reported that a quantitation limit below 0.1 area per cent of the mean compound was achievable with 100 $\mu$m capillaries. Again it was found that impurities not previously separated by HPLC were effortlessly separated by CEC. Demarest's group have compared the effectiveness of using CEC to assess the impurity profile of a neutral and very hydrophobic pharmaceutical drug substance and have shown that the accuracy and precision of the CEC method was acceptable from a regulatory position.[51,52] Norgestimate and its impurities are very difficult to separate by gradient-HPLC owing to their differing hydrophobicities, hence a CEC method has been developed.[53] Isoelutropic theory was used to calculate a binary organic modifier system with the same solvent strength as achieved with gradient-HPLC. The separation was found to be achieved in less than 15 min which was less than half the analysis time required for HPLC. The method was found to be able to quantitate 0.1% degradation impurities in the norgestimate drug substance. The fact that CEC produces Gaussian-shaped peaks of high efficiency means that improved detection levels should be feasible, especially for small peaks on the tailing side of peaks. The downside of capillary-based separations resides in the low concentration sensitivity of the technique. However, numerous ways of circumventing the problem exist and they will be addressed in the following section.

## UV Detection

A recognized problem with all capillary separation techniques is the low concentration sensitivity with UV detection. In order to overcome the problem of low sensitivity of CEC several approaches have been fabricated.

The most commonly used solution is the use of a high-sensitivity UV detection cell (Z-flow cell), this cell increases the detection path length thus allowing the sensitivity of the technique to be increased. This approach has proved to be very successful and several workers have reported that an increase in signal-to-noise ratio of up to 10:1 could be achieved with almost no loss of resolution.[41,54]

An alternative approach is to focus the analyte band on the top of the capillary by dissolving the analyte in a weaker sample solvent than the mobile phase.[31]

From the examples reported to date, it is apparent that CEC should allow the pharmaceutical industry to detect impurities at the required levels set by regulatory authorities either alone or with the techniques of on-column focusing and/or the use of Z-flow cells.

# Quantitative Assays of Pharmaceutical Products

The CEC analysis of model neutral compounds has resulted in excellent repeatabilities,[5] <0.2% for retention time and <1.5% for peak area, which are acceptable for current pharmaceutical standards. In contrast, subsequent workers have found that the repeatability of the analysis was unacceptable unless an internal standard was employed (see Table 7.2). The peak area %RSD in CEC is dependent on the analyte load, which, in turn, is dependent on the sample concentration and whether an electrokinetic (and hence the magnitude of the EOF) or hydrodynamic injection was used. Therefore, extreme care has to be taken in comparing results. The mixed mode phase with its increased EOF may be of particular advantage here as higher analyte loadings are possible.[17]

## Gradient CEC

At present, commercially available CE equipment which is capable of performing CEC is limited to simple isocratic CEC experiments. A significant drawback is its inability to perform gradient elution CEC. In order to realise the full potential of CEC, it is necessary to develop the capacity of gradient elution for the separation of complex mixtures of widely differing lipophilicities as in HPLC. It is apparent from Figure 7.4(a) that the six diuretics possess widely differing log $P$ values and that a gradient CEC would be preferred. Commercially available CEC systems will allow automated step-gradients[39] to be performed as can be seen in Figure 7.4(b). However, from a pharmaceutical view point (unacceptable baseline disturbances would limit the applicability of the approach for pharmaceutical impurity screening) it would be advantageous to perform continuous gradient CEC.

There have been several publications in the literature of in-house built gradient CEC systems,[37,38,55,56] in which the CEC capillary takes the changing mobile phase composition on demand, *i.e.* there is no pressure flow down the CEC capillary so plug flow is maintained. Preliminary investigations have been performed into the use of continuous gradient CEC using a modified HP$^{3D}$ CE system.[56]

The mobile phase was supplied to the capillary vial *via* a binary gradient HPLC pump and degasser. As this liquid stream passed the capillary tip the proprietary design of the inlet electrode and vial allowed the capillary to take the mobile phase on demand by the action of the EOF.

**Table 7.2** *Reproducibility of CEC for the analysis of pharmaceuticals using an internal standard*

| Compound | RSD for retention time % | RSD for peak area % | Reference |
|----------|--------------------------|---------------------|-----------|
| Cannabinoids | <0.1 | <5.0 | 41 |
| Norgestimate | <2.0 | <2.0 | 53 |
| Nicotinamide | 1.8 | 0.63 | 58 |

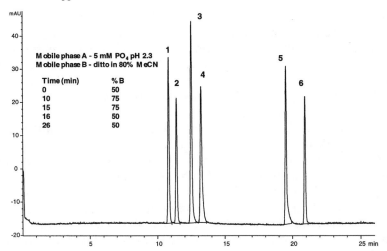

**Figure 7.8** *CEC separation of the diuretics chlorothiazide (1), hydrochlorothiazde (2), chlorothalidone (3), hydroflumethiazide (4), bendroflumethiazide (5) and bumetanide (6). Electrochromatography performed on a Waters Spherisorb ODS 1 column (25 cm, 50 µm i.d.)*

The repeatability of the gradient composition was found to be excellent for a range of gradient shapes, including simple linear, concave, convex and linear containing isocratic steps. Typically RSDs below 1% were observed.

Using the continuous gradient system the separation of the same diuretic mixture on a Waters Spherisorb ODS 1 phase was achieved in a reduced analysis time and gave sharper peaks for the later eluting peaks than could be achieved for isocratic elution alone, see Figures 7.4(a) and (b) and 7.8. The evaluation of the prototype instrumentation gave extremely encouraging results, in that the concept was found to be practicable and repeatable and that pharmaceutically relevant compounds with widely differing lipophilicities could be analysed by gradient CEC.

## Rapid-analysis CEC

The use of a short-end injection technique with reverse polarity has been reported for the rapid analysis of extremely lipophilic steroids.[57] This technique facilitated the separation of budesonide and related steroids in approximately 1 min. The attractiveness of this approach resides in the fact that most of the voltage drop occurs over the short packed capillary (approximately 7.5 cm) rather than over the entire capillary. Therefore higher EOF values are obtainable and hence higher loadings.

## 5 The Future of CEC in the Pharmaceutical Industry

CEC has been proven to be a very powerful separation technique with a higher resolving power than conventional HPLC. When analytes are run in their ionized form the technique is orthogonal to both HPLC and CE. However, if CEC is to be integrated into the pharmaceutical industry on a routine basis, a number of issues will need to be resolved. CEC needs to be able to provide robust, quantitative methods of analysis with the ability to detect low-level impurities in pharmaceuticals of wide functionality which are comparable with, or better than, the methods of analysis already in place.

In order to address this, the pharmaceutical industry eagerly awaits a new and purpose-designed instrumentation which would be capable of performing isocratic and gradient CEC and micro-LC and voltage-assisted micro-LC coupled to high-sensitivity detectors (both UV and MS). In addition, one must not forget the stationary phase which, after all, is the heart of the CEC system, providing both the separation selectivity and the flow. Significant advances in stationary phase chemistry must be made – *i.e.* we desperately require specifically designed phases which can control and modulate the EOF over a wide range of pH values. In addition, a reappraisal of the packed capillary design is needed, as there is still a perception that packed capillaries are fragile and demanding in the hands of an inexperienced operator. It is fully expected that CEC will become a major separation technique by 2010 as rapid developments in the field take place.

# References

1. N.W. Smith and M.B. Evans, *Chromatographia*, 1995, **41**, 197.
2. C. Yan, D. Schaufelberger, and F. Erni, *J. Chromatogr. A*, 1994, **670**, 15.
3. R.J. Boughtflower, T. Underwood, and J. Maddin, *Chromatographia*, 1995, **41**, 398.
4. R.J. Boughtflower, T. Underwood, and C.J. Paterson, *Chromatographia*, 1995, **40**, 329.
5. M.M. Robson, S. Roulin, S.M. Shariff, M.W. Raynore, K.D. Bartle, A.A. Clifford, P. Myers, M.R. Euerby, and C.M. Johnson, *Chromatographia*, 1996, **43**, 313.
6. L.A. Frame, M.F. Robinson, and W.J. Lough, *J. Chromatogr. A*, 1999, **798**, 243.
7. G. Ross, M. Dittmann, F. Bek, and G. Rozing, *Am. Lab.*, 1995, 34.
8. M.M. Dittmann and G.P. Rozing, *J. Chromatogr. A*, 1994, **744**, 63.
9. J.H. Knox and I.H. Grant, *Chromatographia*, 1987, **24**, 135.
10. J.H. Knox and I.H. Grant, *Chromatographia*, 1991, **32**, 32.
11. M.M. Dittmann, K. Wienand, K. Bek, and G.P. Rozing, *LC-GC*, 1995, **13**, 802.
12. M.R. Euerby, C.M. Johnson, and K.D. Bartle, *LC-GC Int.*, 1998, **11**, 39.
13. M.R. Euerby, presented at the Royal Society of Chemistry, Analytical Division Meeting, New Development and Applications in Electrochemistry, Bradford, UK, December, 1997.
14. I.S. Lurie, presented at HPCE 98, Orlando, FL., February, 1998.
15. N.C. Gillott, M.R. Euerby, C.M. Johnson, D.A. Barrett, and P.N. Shaw, *Anal. Commun.*, 1998, **35**, 217.
16. M.G. Cikalo, K.D. Bartle, M.M. Robson, P. Myers, and M.R. Euerby, *Analyst*, 1998, **123**, 87R.
17. M.R. Euerby and C.M. Johnson, personal communication.
18. M.R. Euerby, D. Gilligan, C.M. Johnson, S.C.P. Roulin, P. Myers, and K.D. Bartle, *J. Microcol. Sep.*, 1997, **9**, 373.

19. N.W. Smith, presented at the 1st International Symposium on Capillary Electro-chromatography, San Francisco, CA, August 1997.
20. M.M. Dittmann, presented at the 18th International Symposium on Capillary Chromatography, Riva del Garda, Italy, May 1996.
21. N.W. Smith and M.B. Evans, *Chromatographia*, 1994, **38**, 649.
22. S.J. Lane, R. Boughtflower, C. Paterson, and T. Underwood, *Rapid Commun. Mass Spectrom.*, 1995, **9**, 1283.
23. F. Lelieve, C. Yan, R.N. Zare, and P. Gareil, *J. Chromatogr. A*, 1996, **723**, 145.
24. S.J. Lane, R.J. Boughtflower, C.J. Paterson, and M. Morris, *Rapid Commun. Mass Spectrom.*, 1996, **10**, 733.
25. D.B. Gordon, G.A. Lord, and D.S. Jones, *Rapid Commun. Mass Spectrom.*, 1994, **8**, 544.
26. M.R. Euerby, C.M. Johnson, S.C.P. Roulin, P. Myers, and K.D. Bartle, *Anal. Commun.*, 1996, **33**, 403.
27. K. Schmeer, B. Behnke, and E. Bayer, *Anal. Chem.*, 1995, **67**, 3656.
28. J.H. Miyawa, M.S. Alasandro, and C.M. Riley, *J. Chromatogr. A*, 1997, **769**, 145.
29. J.H. Miyawa, D.K. Lloyd, and M.S. Alasandro, *J. High. Resolut. Chromatogr.*, 1998, **21**, 161.
30. J.H. Miyawa, D.K. Lloyd, and M.S. Alasandro, presented at 1st International Symposium on Capillary Electrochromatography, San Francisco, CA, August 1997.
31. D.A. Stead, R.G. Reid, and R.B. Taylor, *J. Chromatogr. A*, 1998, **798**, 259.
32. G.A. Lord, D.B. Gordon, P. Myers, and B.W. King, *J. Chromatogr. A*, 1997, **768**, 9.
33. R.M. Seifer, W.Th. Kok, J.C. Kraak, and H. Poppe, *Chromatographia*, 1997, **46**, 131.
34. G. Ross, M.M. Dittmann, and G.P. Rozing, presented at the 19th International Symposium on Capillary Chromatography and Electrophoresis, Wintergreen, VA, May 1997.
35. D.A. Stead, R.G. Reid, and R.B. Taylor, presented at the 21st International Symposium on High Performance Liquid Phase Separations and Related Techniques, Birmingham, UK, June 1997.
36. J.C. Kraak, S. Heemstra, R. Swart, and H. Poppe, presented at the 19th International Symposium on Capillary Chromatography and Electrophoresis, Wintergreen, VA, May 1997.
37. C. Huber, G. Choudhary, and C. Horváth, *Anal. Chem.*, 1997, **69**, 4429.
38. M.R. Taylor and P. Teale, *J. Chromatogr. A*, 1997, **768**, 89.
39. M.R. Euerby, D. Gilligan, C.M. Johnson, and K.D. Bartle, *Analyst*, 1997, **122**, 1087.
40. M.R. Euerby, C.M. Johnson, S.F. Smyth, N.C. Gillott, D.A. Barrett, and P.N. Shaw, *J. Microcol. Sep.*, 1999, **11**, 305.
41. I.S. Lurie, R.P. Meyers, and T.S. Conver, *Anal. Chem.*, 1998, **70**, 3255.
42. G. Luo, W. Wei, P. Hu, Y. Wang, R. Wang, and C. Yan, presented at the 9th International Symposium on High Performance Capillary Electrophoresis and Related Microscale Techniques, Anaheim, CA, January 1997.
43. G.A. Lord, D.B. Gordon, and P. Myers, presented at the 21st Internatioal Symposium on High Performance Liquid Phase Separations and Related Techniques, Birmingham, UK, June 1997.
44. N.W. Smith, presented at the 9th International Symposium on High Performance Capillary Electrophoresis and Related Microscale Techniques, Anaheim, CA, January 1997.
45. P.D. Angus, J.F. Stobaugh, C.W. Demarest, K.M. Payne, K.R. Sedo, L.Y. Kwok, and T. Catalano, presented at the 19th International Symposium on Capillary Chromatography and Electrophoresis, Wintergreen, VA, May 1997.

46. C.W. Demarest, K.M. Payne, K.R. Sedo, E. Victorino, T. Catalano, P. Angus, and J. Stobaugh, presented at the 1st International Symposium on Capillary Electro-chromatography, San Francisco, CA, August 1997.
47. T. Adam and M. Kramer, presented at HPCE 98, Orlando, FL, February 1998.
48. P. Sandra, A. Dermaux, V. Ferraz, M.M. Dittmann, and G. Rozing, *J. Microcol. Sep.*, 1997, **9**, 409.
49. Notes for Guidance on Validation of Analytical Procedures: Methodology, CPMP/ICH/281/95 – step 4, Consensus Guideline, November 1996.
50. A. Carlsson, P. Peterson, and A. Walhagen, presented at the 20th International Symposium on Capillary Chromatography, Riva del Garda, Italy, May 1998.
51. C.W. Demarest, K.M. Payne, K.R. Sedo, E. Victorino, T. Catalano, P. Angus, and J. Stobaugh, presented at the 1st International Symposium on Capillary Electro-chromatography, San Francisco, CA, August 1997.
52. K.M. Payne, C.W. Demarest, K.R. Sedo, L.Y. Kwok, and T. Catalano, presented at the 19th International Symposium on Capillary Chromatography and Electro-phoresis, Wintergreen, VA, May 1997.
53. J. Wang, D.E. Scaufelberber, and N.A. Guzman, *J. Chromatogr. Sci.*, 1998, **36**, 155.
54. M.R. Euerby, presented at the 19th International Symposium on Capillary Chromatography and Electrophoresis, Wintergreen, VA, May 1997.
55. C. Yan, R. Dadoo, R.N. Zare, D.J. Rakestraw, and D.S. Anex, *Anal. Chem.*, 1996, **68**, 2726.
56. M.R. Euerby, A.M. Germon, C.M. Johnson, W. Eberhardt, G. Ross, K. Witt, and F. Bek, presented at the 20th International Symposium on Capillary Chromato-graphy, Riva del Garda, Italy, May 1998.
57. M.R. Euerby, C.M. Johnson, M. Cikalo, and K.D. Bartle, *Chromatographia*, 1998, **47**, 135.
58. K. Altria, presented at HPCE 98, Orlando, FL, February 1998.

CHAPTER 8

# Capillary Electrochromatography in Natural Product Research

AN DERMAUX AND PAT SANDRA

## 1 Introduction

It is now generally recognized that the newest separation technology, capillary electrochromatography (CEC), has great potential and may replace high-performance liquid chromatography (HPLC) and complement capillary electrophoresis (CE) for separations requiring high resolution and high peak capacity. The main advantages of CEC over HPLC and CE and the present shortcomings of the neophyte technique have been discussed in other chapters of this book. It is our strong belief, however, that the various technological hurdles will be overcome in the next decade and that more and more real-world applications will demonstrate the potential of CEC.

Working in natural product research means that we are often confronted with multicomponent analysis, so we have carefully followed the evolution and development of CEC and have initiated research into the analysis of natural products with a 'hydrophobic' character. This chapter summarizes our investigations of the last three years.

## 2 Experimental

A Hewlett-Packard HP$^{3D}$CE system (Agilent Technologies, Waldbronn, Germany) equipped with diode array detection (DAD) was used throughout this study. The buffer vials at both the inlet and outlet side were pressurized at 10 bar using nitrogen. Fused silica capillary columns of approximately 35 cm and 50 cm in length with 0.350 mm o.d. and 0.100 mm i.d. (Polymicro Technologies, Phoenix, Arizona) were slurry packed with 3 $\mu$m Hypersil ODS particles (Shandon HPLC, Cheshire, UK) according to the procedures described.[1] The lengths of the packed beds were 25 cm and 40 cm, respectively.

All experiments were performed using on-capillary DAD-UV detection. Sample preparation and other experimental conditions are specified for each application.

# 3 Triglycerides

## Introduction

Triglycerides are the most abundant constituents of oils and fats. They consist of a glycerol molecule in which each hydroxy group has been esterified with a fatty acid. This results in an intricate series of compounds and highly efficient separation techniques are required to unravel this complexity. Triglycerides are simply named by the fatty acid composition; for example PPO stands for a triglyceride containing two palmitic and one oleic ester chains. The most important fatty acids occurring in oils and fats are listed in Table 8.1.

The separation of triglycerides has been extensively studied by capillary gas chromatography (cGC),[2–4] liquid chromatography (LC)[5–9] and supercritical fluid chromatography (SFC).[10–12] Nowadays the primary methods for triglyceride analysis are reversed-phase LC (RPLC)[5] and silver ion chromatography (SIC).[8] The former provides separations according to hydrophobicity, while the latter gives separations according to degree of unsaturation. Micropacked and microbore LC columns were successfully used in our research activities on the triglyceride composition of vegetable oils,[13,14] and it seemed obvious to evaluate CEC, providing higher efficiencies, for the same solutes and samples.

## Separation Principle

In a reversed-phase mechanism, which was applied in CEC, the separation of triglycerides takes place according to the partition number (PN) which is defined as $PN = CN - 2NDB$ in which CN is the carbon number and NDB is the number of double bonds. In the past, compounds presenting the same PN

**Table 8.1** *Important fatty acids occurring in oils and fats*

| Systematic name | Symbols | |
| --- | --- | --- |
| | *Chain composition*[a] | *Abbreviation*[b] |
| Tetradecanoic acid | C14:0 | M |
| Hexadecanoic acid | C16:0 | P |
| Hexadecanoic acid | C16:1 | Po |
| Octadecenoic acid | C18:0 | S |
| Octadecenoic acid | C18:1 | O |
| Octadecadienoic acid | C18:2 | L |
| Octadecatrienoic acid | C18:3 | Ln |
| Octadecatetraenoic acid | C18:4 | Lnn |
| Eicosanoic acid | C20:0 | E |
| Eicosenoic acid | C20:1 | Eo |
| Eicosapentaenoic acid | C20:5 | E5 |
| Docosahexaenoic acid | C22:6 | D6 |

[a]C18:2 refers to a fatty acid with 18 carbon atoms and two double bonds. All natural fatty acids have the *cis* configuration. [b]Triglycerides are indicated by their fatty acid content e.g. POP and LLL.

value, *e.g.* OLL and PLL, were considered as critical pairs and tended to elute close together. With the utilization of more efficient LC columns, many critical PN pairs can nowadays easily be separated and for a specific PN number the lipids with the highest content of saturated fatty acid exhibit the longest retention *e.g.* retention time of PLL > OLL. However, highly unsaturated triglycerides with the same PN number, *e.g.* OLLn and LLL, are still critical pairs in reversed-phase LC.

In a first attempt to analyse triglycerides by CEC, 50 mM ammonium acetate (98% from Sigma Alldrich, Steinheim, Germany) was added to the mobile phase optimized for isocratic elution by micro-LC and consisting of acetonitrile/isopropanol/*n*-hexane in a ratio 57/38/5.[15] A primrose oil sample was analysed with a voltage of 30 kV at 20 °C on a 25-cm column. Injection was performed hydrodynamically at 10 bar during 3 s. Figure 8.1 compares the chromatograms for the same sample obtained by micro-LC on a 50 cm × 0.32 mm i.d. FSOT column packed with 5 µm BioSil C18 HL (A) and on the CEC column (B). The chromatograms were recorded at 210 nm (A) and at 200 nm (B), respectively. The separation profile is, as expected, similar but

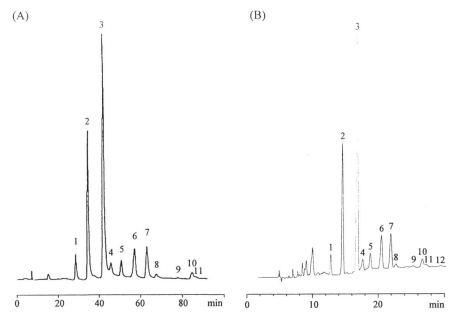

**Figure 8.1** *Triglyceride analysis of primrose oil. (A) Micro-LC: column, 50 cm × 0.320 mm i.d., FSOT, Biosil C18 HL 5 µm. Mobile phase, acetonitrile/isopropanol/ n-hexane 57/38/5 at 5 µl min⁻¹. Detection, UV at 210 nm. T 20 °C. (B) CEC: column, 25 cm × 0.1 mm i.d., FSOT, Hypersil ODS 3 µm. Mobile phase, acetonitrile/isopropanol/n-hexane 57/38/5 – 50 mM ammonium acetate. Detection UV at 200 nm. T 20 °C; voltage, 30 kV; injection 10 bar during 3 s. Peaks: 1, LγLnLn; 2, LLγLn; 3, LLL; 4, OγLnL; 5, PγLnL; 6, OLL; 7, PLL; 8, PγLnO; 9, OLO; 10, SLL; 11, PLO; 12, PLP*

with improved resolution (*e.g.* LLL/OγLnL – peaks 3 and 4) and in a shorter analysis time (factor 3).

Note that this application involves non-aqueous CEC and that the origin of the chemicals used played an important role in obtaining and reproducing these results. In particular, the origin of the ammonium acetate was critical. In fact with ammonium acetate from another supplier difficulties were encountered in solubilizing the salt in the non-aqueous mobile phase. Solubility with other batches from the same supplier, on the other hand, offered no problems.

## Influence of Applied Voltage

The influence of the voltage (15, 20, 25 and 30 kV) on the current, the mobile phase velocity, the efficiency and the resolution was evaluated for arachide (peanut) oil on a 25 cm column at 20 °C (Figure 8.2). For the selected mobile phase composition, the current ranged between 3.84 $\mu$A (15 kV) and 7.57 $\mu$A (30 kV) and the electroosmotic flow (EOF) between 0.36 mm s$^{-1}$ (15 kV) and 0.73 mm s$^{-1}$ (30 kV). The plots of current and EOF *versus* voltage showed a correlation coefficient of 0.999, which indicates that the influence of Joule heating is negligible. Moreover, during analysis, the currents remained very stable. The plate number and resolution reached a maximum at 20 kV (Figure 8.3). At this voltage a current of 5.12 $\mu$A and a mobile phase flow of 0.48 mm s$^{-1}$ was obtained and the reduced plate height ($h$) for the triglyceride POP ($k = 4.3$) was 2.07. This is about 25% higher than the $h$-value of 1.52

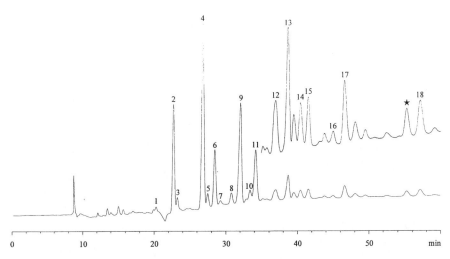

**Figure 8.2**   *Triglyceride analysis of arachide oil. Column and conditions as in Figure 8.1(B) except voltage at 20 kV. Insert: from 35 min enlarged by a factor of 5. Peaks: 1, LLLn; 2, LLL; 3, OLnL; 4, OLL; 5, OLnO; 6, PLL; 7, POLn; 8, PLnP; 9, OLO; 10, SLL; 11, PLO; 12, PLP; 13, OOO; 14, SLO; 15, POO; 16, PLS, 17, POP; 18, POS*

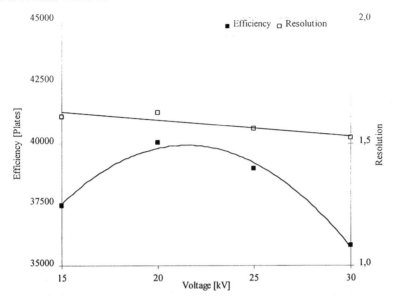

**Figure 8.3** *Efficiency (peak 17 POP) and resolution (peaks ★ POS) as a function of the applied voltage. Column and conditions see Figure 8.2*

calculated on the same column for a polyaromatic hydrocarbon eluting with $k = 4.2$ with a mobile phase composed of acetonitrile/TRIS-HCl (80/20) at pH 8. It is important to note the slow slope of the resolution curve at voltages higher than optimal. Voltages of 25 and 30 kV can be applied without important loss in resolution. The analysis times are, however, drastically reduced by applying higher voltages; *i.e.* by 23% and 39% for 25 and 30 kV, respectively. For longer columns, *e.g.* 40-cm column, 30 kV is necessary to provide acceptable analysis times.

## Influence of Column Length

Figure 8.4(B) shows the separation of corn oil to illustrate the enhanced performance using a 40 cm column. The oil sample (50 mg ml$^{-1}$ isopropanol) was analysed with a voltage of 30 kV at 20 °C. Injection was performed electrokinetically at 10 kV during 3 s. The chromatogram was recorded at 200 nm. The mobile phase velocity was 0.47 mm s$^{-1}$ and 71 000 plates were calculated for POP which represents a reduced plate height of 1.88. The separation is compared to the micro-LC analysis of the same sample [Figure 8.4(A)]. The micro-LC analysis was performed on a $2 \times 25$ cm $\times$ 1 mm i.d. stainless-steel (SS) column packed with 5 $\mu$m Biosil C18 HL using evaporative light-scattering detection. The superior performance of CEC compared to LC, and this in roughly the same analysis time, is obvious.

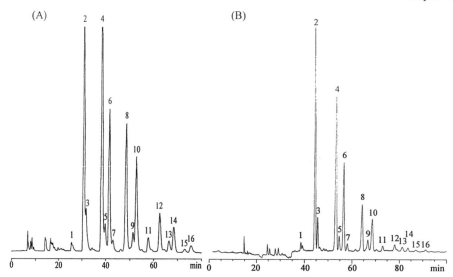

**Figure 8.4**  *Triglyceride analysis of corn oil. (A) Micro-LC: column, 2 × 25 cm × 1 mm
i.d., stainless steel (SS), Biosil C18 HL 5 μm. Mobile phase, acetonitrile/
isopropanol/n-hexane 57/38/5 at 55 μl min⁻¹. Detection: evaporative light-
scattering detection (ELSD) (Sedex 45) at 50 °C, 1.8 bar nitrogen, range 4.
T' 20 °C. (B) CEC: column, 40 cm × 0.1 mm i.d., FSOT, Hypersil ODS 3 μm.
Mobile phase, acetonitrile/isopropanol/n-hexane 57/38/5 – 50 mM ammonium
acetate. Detection UV at 200 nm. T' 20 °C; voltage 30 kV; injection 10 kV
during 3 s. Peaks: 1, LLLn; 2, LLL; 3, OLnL; 4, OLL; 5, OLnO; 6, PLL; 7,
POLn; 8, OLO; 9, SLL; 10, PLO; 11, PLP; 12, OOO; 13, SLO; 14, POO;
15, PLS; 16, POP*

## Influence of Column Temperature

The influence of the column temperature on the separation was investigated in
the 10 °C to 50 °C range with 10° intervals for the CEC analysis of sunflower oil
at 30 kV on a 25 cm column. The results are shown in Table 8.2. At 50 °C, peaks
were broad and even distorted. Although 10 °C gives the best results, 20 °C is
favoured for practical applications because efficiency and resolution are acceptable, analysis times are shorter and stabilization of the temperature is easy.

**Table 8.2**  *The influence of column temperature on reduced plate
height and mobile phase velocity*

| Temperature (°C) | Reduced plate height (h) | Mobile phase velocity (u)/mm s⁻¹ |
|---|---|---|
| 10 | 1·95 | 0·63 |
| 20 | 2·3 | 0·74 |
| 30 | 2·6 | 0·82 |
| 40 | 2·9 | 0·92 |

## Influence of Injection Mode

The injection of lipid samples may be performed electrokinetically or hydro-dynamically by pressure. No differences were observed in the qualitative and quantitative profiles of the samples applying 10 kV or 10 bar during 3 s, which can be explained by the neutral character of the triglycerides. Electrokinetic injection, however, gives better reproducibility of migration time data. A salad oil composed of a mixture of 80% sunflower oil and 20% olive oil was analysed to investigate the reproducibility of both electrokinetic and hydro-dynamic injection. Relative standard deviations (RSDs) on migration times of the eight major triglycerides for 10 consecutive runs were substantially lower for electrokinetic injection (%RSD ≤ 0.45) compared to hydrodynamic injection (%RSD ≤ 1.32). This confirms data on hydrodynamic and electrokinetic injection in CEC described by Ross *et al.*[16] Optimized injection time for the analysis of triglycerides is 3 s or shorter. Injection during longer periods with isopropanol as solvent resulted in severe band broadening, as illustrated in Figure 8.5 for the analysis of the triglycerides extracted from soya lecithin.

Compared to the other samples, the concentration of the lipids was lower in the injection vial ($\approx 10$ mg ml$^{-1}$) and the injection time was increased to 12 s. The efficiency compared to a 3 s injection dropped by 25%. Nevertheless, the separation was still much better than the one of soya oil presented in the literature on $2 \times 25$ cm $\times$ 4.6 mm i.d. columns packed with 5 $\mu$m ODS particles.[7]

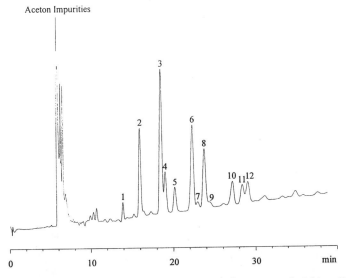

**Figure 8.5** *Analysis of the triglycerides extracted from soya lecithin. Column and conditions as in Figure 8.1(B) except injection during 12 s. Peaks: 1, LLnLn; 2, LLLn; 3, LLL; 4, OLnL; 5, PLnL; 6, OLL; 7, OLnO; 8, PLL; 9, POLn; 10, OLO; 11, SLL; 12, PLO*

# Applications

CEC has been applied to analyse more than 30 samples (vegetable oils, margarines, pharmaceutical formulations, *etc.*) and in all cases more complete triglyceride profiles were obtained with CEC than with LC.[14,15]

## *Analysis of a Pharmaceutical Formulation*

A pharmaceutical application is shown in Figure 8.6 for the analysis of a formulation consisting of some testosterone esters and a lipid mixture. The analysis was performed on a 25 cm column using acetonitrile/isopropanol/$n$-hexane in the ratio 57/38/5 – 50 mM ammonium acetate. The sample (50 mg ml$^{-1}$ isopropanol) was analysed with a voltage of 30 kV at 20 °C. Injection was performed hydrodynamically at 10 bar during 1.2 s. The less hydrophobic testosterone esters elute before the lipids and were detected from 0 to 9 min at 254 nm. The elution order of the testosterone esters could be elucidated by the recorded DAD spectra. The first peak corresponds to testosterone phenylpropionate (peak 1') while the others are testosterone propionate (peak 2'), testosterone isocaproate (peak 3') and testosterone decanoate (peak 4'). The lipid profile detected at 200 nm corresponds to peanut oil.

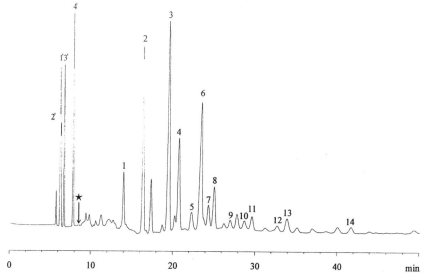

**Figure 8.6**   *Analysis of a pharmaceutical formulation. Column and conditions as in Figure 8.1(B) except detection from 9–9 min at 254 nm (★. Peaks: 1', testosterone phenylpropionate; 2', testosterone propionate; 3', testosterone isocaproate; 4', testosterone decanoate; 1, LLLn; 2, LLL; 3, OLL; 4, PLL; 5, PLnP; 6, OLO; 7, SLL; 8, PLO; 9, OOO; 10, SLO; 11, POO; 12, PLS; 13, POP; 14, POS*

## Analysis of Fish Oil

Vegetable oils are relatively simple in their composition because the common fatty acids are P, S, O, L and Ln. Fish oils on the other hand, are characterized by a broader range of carbon numbers and a higher degree of unsaturation. This is obvious from the CEC analysis of a sardine oil sample shown in Figure 8.7.

The same experimental conditions as for Figure 8.4(B) were applied except that the solution was 150 mg in 1 ml isopropanol. More than 100 peaks are registered and it is impossible to elucidate the composition of the triglycerides by a one-dimensional separation. Nevertheless, the high resolving power of CEC is well illustrated and, as far as we know, this is the most complete separation of the triglycerides in fish oil ever demonstrated.[17]

Fish oils are interesting to study because of their therapeutic value and their occurrence in vitamin formulations. Because of their hypertriglyceridemiant action they have beneficial effects in the treatment of atheroma and vascular thrombosis. Medical studies[18–20] have demonstrated that the polyunsaturated fatty acids from the $\bar{\omega}$-3 series (*n*-3), especially eicosanepentaenoic acid C20:5(*n*-3) and docosanehexaenoic acid C22:6(*n*-3) are the active principles with metabolic action in the constitution of cellular membranes and synthesis of prostaglandins, thromboxanes and leucotrienes.

In-depth characterization of fish oil requires a multimodal or multidimensional approach, so a novel approach was used for the identification of triglycerides in fish oils. Prefractionation was performed by silver ion packed column supercritical fluid chromatography (SI-pSFC) and the collected fractions were further analysed by RP-CEC.[21] The separation of triglycerides in vegetable oils using both SI-pSFC and RP-pSFC has recently been discussed.[22–26] Reproducible results in SI-pSFC were obtained using modifier and pressure

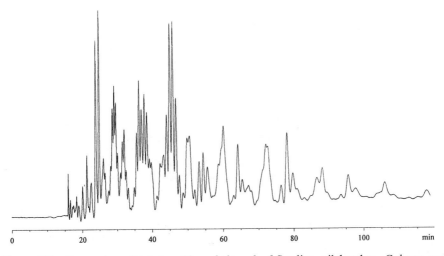

**Figure 8.7** *Separation of triglycerides of the oil of* Sardina pilchardus. *Column and conditions as described in Figure 8.4(B)*

gradients at constant pressure. Eighteen fractions were collected and analysed by CEC and direct injection in electrospray mass spectroscopy (ESMS). The data illustrated that the fraction numbers corresponded to the number of double bonds in the lipids, *e.g.* fraction 1 contains 1 double bond, fractions 11 and 18 contain 11 and 18 double bonds, respectively. This means that the lipids in each fraction only differ in carbon number and not in degree of unsaturation. The last fraction (18, thus 18 double bonds) contains only one triglyceride namely the lipid with three C22:6 fatty acids.

Figure 8.8 illustrates the CEC (A) and ESMS (B) analysis of the fraction containing 10 double bonds. This approach allowed us to elucidate the lipids in that fraction, which are listed in Table 8.3.

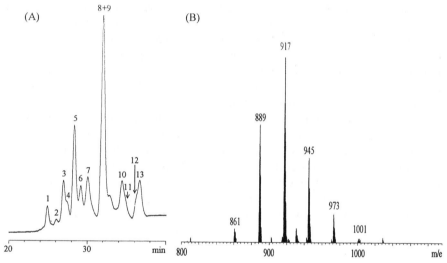

**Figure 8.8** *Analysis of triglycerides in SI-SFC fraction 10 (see text). (A) CEC: column and conditions as described in Figure 8.4(B). Peak analysis is shown in Table 3.3. (B) ESMS: instrument: HP5939B quadrupole mass spectrometer and HP 59987A atmospheric pressure ionization (API) electrospray source (Hewlett-Packard, Palo Alto, CA, USA). Positive ion mode. Fragmentor at 25 V; injection, 1 μl of fraction*

## *Fraction Collection Followed by Off-line ESMS*

An interesting feature of the CE instrument used is the possibility of collecting fractions or peaks. The software package includes a dialog box (HPCE Fraction Collection) for automatic collection *via* time windows specified by the migration time and the collection width. Both values are easily accessible for electrophoretic and/or micellar electrokinetic operation but the situation is more complicated for CEC operation. The residence time of a solute in the total capillary is the sum of its migration time in the packed bed and the migration time in the empty part (detector vial) of the capillary. The first value equals its

**Table 8.3** *Detailed analysis of fraction 10 from the separation of triglycerides from vegetable oils using SI-pSFC. See text for details*

| Peak notation[a] | MS signal | Carbon atoms | Partition number (PN) | Triglyceride structure assignment | Retention interval (according to PN) | |
|---|---|---|---|---|---|---|
| | | | | | $t_i$/min | $t_f$/min |
| 1 | 861 | 52 | 32 | 20:5/18:4/14:1 | 24 | 26 |
| 2 | 889 | 54 | 34 | 20:5/18:4/16:1 | 26 | |
| 3 | | | | 22:6/18:4/14:0 | | |
| 4 | | | | 20:5/20:5/14:0 | | 28 |
| 5 | 917 | 56 | 36 | 20:5/18:4/18:1 | 28 | |
| 6 | | | | 20:5/20:5/16:0 | | |
| 7 | | | | 22:6/18:4/16:0 | | 32 |
| 8 | 945 | 58 | 38 | 20:5/20:5/18:0 | 32 | |
| 9 | | | | 22:6/18:4/18:0 | | 33·5 |
| 10 | 973 | 60 | 40 | 20:5/20:5/20:0 | 33·5 | |
| 11 | | | | 22:6/18:4/20:0 | | 36 |
| 12 | 1001 | 62 | 42 | 22:5/20:5/22:0 | 36 | |
| 13 | | | | 22:6/22:0/18:4 | | 38 |

[a]Peak notation from Figure 8.8.

detection time while the second value should be deduced from the empty column length and the EOF. The principle was illustrated by Sandra *et al.*[15] for the collection of OOO in salad oil. OOO detected at 39.402 min has migrated in the packed bed at a velocity of 0.111 mm s$^{-1}$ but in the unpacked part of the column, in this case 8.5 cm, at the EOF of 0.499 mm s$^{-1}$. The time spent in the empty part was therefore 2.837 min. The peak maximum reached the collection vial at 42.239 min. The time window programmed was 41.2 to 43.3 min and OOO was collected 8 times in 100-$\mu l$ mobile phase. The sample was analysed as such by direct ESMS. A pure ESMS spectrum was obtained, which confirmed the identity of OOO, namely by $m/z$ 903 (OOO + NH$_4^+$), $m/z$ 908 (OOO + Na$^+$) and $m/z$ 924 (OOO + K$^+$) (Figure 8.9). The fragment $m/z$ 603 corresponds to OO$^+$.

# 4 Fatty Acids and Their Derivatives

## Introduction

The most important fatty acids naturally occurring in vegetable oils and margarines are palmitic acid (C16:0), stearic acid (C18:0), oleic acid (C18:1), linoleic acid (C18:2), and linolenic acid (C18:3). The number of double bonds is represented by :*x* and they all have the *cis* configuration. Fatty acids with double bonds of *trans* configuration are occasionally found in natural lipids, but are

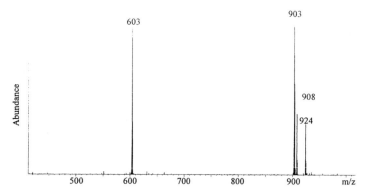

**Figure 8.9**   *Electrospray MS spectrum of OOO collected from salad oil. Conditions as described in Figure 8.8(B) except fragmentor at 250 V*

mainly formed during industrial processing, *e.g.* the production of margarine by hydrogenation of vegetable oils.

The features of CEC for fatty acid analysis were evaluated by analysing the free fatty acids (FFAs) as such as well as the methyl ester (FAME) and phenacyl ester (FAPE) derivatives.[27]

## Preparation of FFAs, FAPEs and FAMEs

To 500 mg oil or fat, 10 ml of a solution of 1 M potassium hydroxide in 95% ethanol was added and the mixture was refluxed for one hour. After cooling to ambient temperature, 25 ml water was added and the non-saponifiable material was removed by extraction with 30 ml diethyl ether. The aqueous layer was then acidified with 1 M hydrochloric acid and the liberated FFAs were extracted with 3 × 15 ml diethyl ether. The combined ether layers were washed with 15 ml water and dried over anhydrous sodium sulfate. The solvent was removed on a rotavapor. Aliquots were taken for direct analysis by CEC or for the preparation of FAPEs and FAMEs. For direct CEC analysis, a 1% solution was made in the mobile phase. For the preparation of the FAPEs, 100 $\mu$l of a phenacyl bromide and triethylamine solution (each 20 mg ml$^{-1}$ in acetone) was added to 50 $\mu$l of a FFA solution (30 mg ml$^{-1}$ in acetone). The vial, fitted with a PTFE cap, was sealed and heated in a boiling water bath for 15 min. After cooling to ambient temperature, 30 $\mu$l acetic acid was added to destroy the excess reagent. The vial was then resealed and heated for 5 min. Finally, the solvents were evaporated in a stream of nitrogen and the phenacyl esters were dissolved in 200 $\mu$l mobile phase. The FAMEs were prepared by adding 1 ml of BF$_3$–methanol (14% w/v) to 10 mg of the FFAs. The vial, fitted with a PTFE cap, was sealed and heated at 100 °C for 2 min. The FAMEs were extracted with 2 ml *n*-hexane. The organic layer was removed under a nitrogen stream and the residue was dissolved in 1 ml mobile phase.

## Separation Principle

In a reversed-phase mechanism, the separation of FFAs, FAMEs and also of FAPEs takes place according to their PN. The separation occurs in increasing order of CN, while one double bond in the chain reduces the retention time by the equivalent of two CNs. Two components with the same PN number are by definition critical pairs (*e.g.* C16:1 and C18:0).

Figure 8.10 compares the isocratic separations of the FAPEs C18:3, C18:2, C18:1, C16:0 and C18:0, obtained by micro-LC on a FSOT column (30 cm × 0.32-mm i.d.) packed with 3 μm ODS particles (RoSil C18 HL from BioRad) with methanol/water (95/5) at a velocity of 0.7 mm s$^{-1}$, and by CEC on the 25 cm and 40 cm columns.

Contrary to our experiments with the CEC and micro-LC analysis of triglycerides, in which the same mobile phase could be applied with exception of the addition of 50 mM ammonium acetate in CEC, the LC mobile phase could not be used for the analysis of FFAs, FAMEs and FAPEs by CEC.

A number of mobile phase compositions were evaluated in CEC and the mixture acetonitrile/50 mM MES pH 6 (90/10 v/v) provided the best separation in terms of selectivity and speed of analysis, but not in efficiency. The FAPEs sample was analysed with a voltage of 30 kV at 20 °C. Injection was performed electrokinetically at 10 kV during 3 s. The chromatograms were recorded at 242 nm.

The CEC columns were first evaluated with some PAHs and the plate numbers measured were 58 000 ($h = 1.4$) and 88 000 ($h = 1.5$) for the 25 cm and 40 cm columns, respectively. For the mobile phase used for the fatty acids and derivatives the plate numbers dropped to 31 000 ($h = 2.7$) and 52 000 ($h = 2.6$). Nevertheless, for the same particle size, CEC performed much better than micro-LC for which the plate number was 28 500 plates ($h = 3.5$) at a velocity of 0.7 mm s$^{-1}$. For the CEC columns the mobile phase velocities were 1.9 mm s$^{-1}$ for the 25 cm and 1.1 mm s$^{-1}$ for the 40 cm column. The currents generated were 5.8 μA and 4.4 μA, respectively.

A better criterion to express the resolving power is the separation number. The $S_N$ between C16:0 and C18:0 is 12.7, 16.0 and 22.4 for the micro-LC, CEC 25-cm and CEC 40-cm column, respectively. The $S_N$ min$^{-1}$ values are 0.20, 0.85 and 0.49, respectively, illustrating the superior performance of CEC compared to micro-LC.

An important selectivity difference was noted between micro-LC and CEC. On the Biorad RoSil phase C16:0 elutes before C18:1 whereas on the Hypersil ODS C18:1 elutes before C16:0. This is due to the difference in mobile phase composition and stationary phase octadecyl loading and residual silanol functions.

## FFA *versus* FAPE Analysis

The main reason for analysing FAPE derivatives instead of the FFAs is the quantitative aspect. The response factors for the different FAPEs are close to 1

(A)

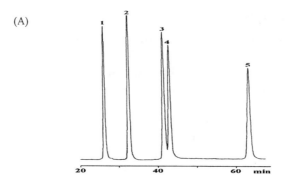

(B)

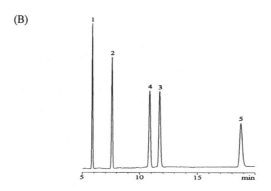

(C)

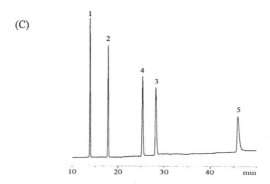

**Figure 8.10** *Separation of FAPEs. (A) Micro-RPLC: column 30 cm × 0.32 mm i.d., FSOT, RoSil C18 HL 3 μm. Mobile phase, methanol/water (95/5 v/v). Injection volume, 60 nl; detection, UV at 242 nm (B) CEC: column, 25 cm × 0.1 mm i.d., FSOT, Hypersil C18 3 μm. Mobile phase, acetonitrile/50 mM MES pH 6 (90/10). Detection, UV at 242 nm. (T' 20 °C; Voltage, 30 kV. Injection, 10 kV during 3 s (C) conditions as described in Figure 8.10(B) except column, 40 cm × 0.1 mm i.d., FSOT, Hypersil C18 3 μm. Peaks: 1, C18:3; 2, C18:2; 3, C16:0; 4, C18:1; 5, C18:0*

so the normalization method can be applied. Moreover the DAD spectra can be used to calculate the number of double bonds in the fatty acid chain.[27] The plot of the ratio of the extinction coefficients at 240 and 210 nm *versus* the number of double bonds allows elucidation of the number of double bonds in unknown samples, *e.g.* fish oils.

Figure 8.11(A) shows a typical CEC separation of the FAPEs from soya oil on the 40 cm column. The FAPEs sample was analysed with a voltage of 30 kV at 20 °C. Injection was performed electrokinetically at 10 kV during 3 s. The chromatogram was recorded at 242 nm. The fatty acids identified are C18:3, C18:2, C18:1, C16:0 and C18:0. Their percentage composition is 6.4, 53.4, 23.2, 13.0 and 4.0, respectively. This corresponds with quantitative data obtained with other techniques, such as capillary GG analysis of the methyl esters.

FFAs and FAMEs can also be separated by CEC. The lack of a suitable chromophore makes UV detection difficult and low wavelengths have to be used. Because of the short detection pathway in electro-driven separation methods, operating at 200 nm or below is feasible. Nevertheless, analysis of FFAs and FAMEs with UV detection makes no sense. On the one hand, the response factors for unsaturated fatty acids C18:3, C18:2 and C18:1 differ too much to be applied for quantitative analysis, and, on the other hand, the saturated fatty acids C16:0 and C18:0 cannot be detected. This discriminative effect is illustrated in Figure 8.11(B) for the FFAs of the soya oil sample, shown in Figure 8.11(A). Note that the migration times of the FFAs are shorter than for the corresponding FAPEs.

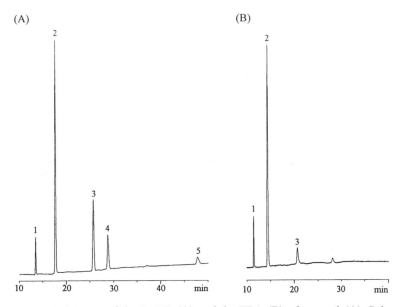

**Figure 8.11** *Separation of the FAPEs (A) and the FFAs (B) of soya oil. (A) Column and conditions as described in Figure 8.10(C). (B) Column and conditions as described in Figure 8.10(C) except detection at 200 nm. Peaks: 1, C18:3, 2, C18:2; 3, C18:1; 4, C16:0; and 5, C18:0*

## Applications

CEC has been applied to determine the fatty acid composition in a large number of oils and margarines, and in all cases more complete fatty acid profiles were obtained with CEC than with LC.[27] From a practical point of view, however, capillary GC is still the method of choice to analyse the fatty acid composition of oils and fats, although margarines may be an exception.

### *Fatty Acid Analysis of Fish Oils*

Figure 8.12 shows the FFAs (A), FAMEs (B) and FAPEs (C) analysis of sardine oil. For the FAPEs the response factors are close to one and the normalization method could be used. The amounts of C18:4, C20:5, C22:6, C16:1, C14:0, C18:1, C16:0, C20:1 and C18:0 are 4.5, 22.5, 20.9, 9.6, 8.7, 10.3, 18.6, 1.6 and 3.3%, respectively.[17] Fish oil is characterized by a high degree of unsaturation, mainly carried by eicosapentaenoic acid (C20:5) and docosahexaenoic acid (C22:6). The DAD spectra could be used to calculate the number of double bonds in the fatty acid moiety of the FAPEs. The absorbance ratios $\varepsilon_{240}/\varepsilon_{210}$ for 6, 5, 4, 3, 2, 1 and 0 double bonds were 0.42, 0.51, 0.64, 0.92, 1.40, 1.83 and 2.06, respectively.

The principle was applied to identify peaks 4 and 5. The ratios were 1.82 and 2.04, respectively, which means that peak 4 is C16:1 and peak 5 is C14:0. This was confirmed for the latter by plotting log $k$ *versus* CN for the saturated fatty acids. The plot is linear which also illustrates that the mobile phase flow created by electroosmosis is constant during the analysis.[17]

### *Determination of C18:1* cis/trans *Ratio in Margarines*

For the analysis of margarines, FFA analysis by CEC gives some interesting information. Hydrogenation of vegetable oils can result in substantial amounts of positional and configurational isomers. Best known is the formation of the *trans* isomer of 9-C18:1, *i.e.* elaidate as well as *cis* and *trans* isomers with the double bonds in 6 and 11 positions. The elucidation of the *cis/trans* ratio of C18:1 is of utmost importance. We normally apply micro-LC on ODS with $Ag^+$ loaded mobile phases after formation of the FAPE derivatives.[14] The same principle could not be applied in CEC because the addition of $Ag^+$ ions resulted in very unstable currents. Some positional and geometrical isomers, however, could be separated by CEC, but this separation is hardly applicable because of overlaps of *cis–trans* isomers and of *trans* isomers with C16:0. To our surprise, the C18:1 *cis/trans* ratio in margarines could be measured in the CEC analysis of the FFAs because C16:0 is not detected at 200 nm.

Figure 8.13 shows the analysis of the FFAs of an experimental margarine, and the *cis* group can nicely be distinguished from the *trans* group. Because both groups have the same response factors the calculated *cis/trans* ratio is 1:3, which was confirmed by micro-LC/ODS/$Ag^+$ of the FAPEs.[28]

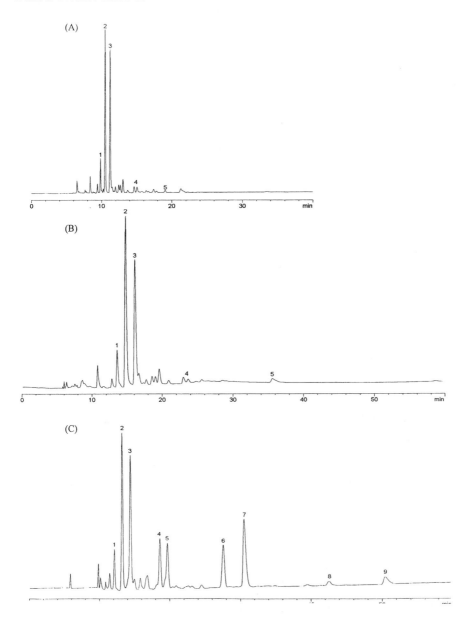

**Figure 8.12** *Separation of free fatty acids* (A), *fatty acid methyl esters* (B) *and fatty acid phenacyl esters of the soil of* Sardina pilchardus. *(A) Column, 40 cm × 0.1 mm i.d., FSOT, Hypersil C18 3 μm. Mobile phase, acetonitrile/50 mM MES pH 6 (90/10). Detection, UV at 200 nm. 20 °C; Voltage, 30 kV; Injection, 10 kV during 3 s.* (B) *Conditions as described in* (A). *Peaks: 1, C18:4; 2, C20:5; 3, C22:6; 4, C16:1; 5, C18:1.* (C) *Conditions as described in* (A) *except UV detection at 242 nm. Peaks: 1, C18:4; 2, C20:5; 3, C22:6; 4, C16:1; 5, C14:0; 6, C18:1; 7, C16:0; 8, C20:1; and 9, C18:0*

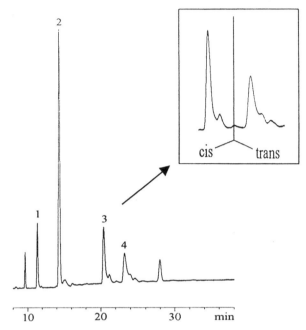

**Figure 8.13** *Separation of FFAs of an experimental margarine. Column and conditions as described in Figure 8.12(A). Peaks: 1, C18:3; 2, C18:2; 3, all-c-C18:1; 4, all-t-C18:1*

# 5 CEC Analysis of Other 'Hydrophobic' Samples

## Introduction

In the last few years CEC has become more and more applied in our laboratory to unravel, by its superior efficiency compared to LC, the complexity of natural products with a hydrophobic character.

Some typical examples have been selected to illustrate the potential of CEC in this respect.

## CEC Analysis of Carotenoids and Tocopherols (Vitamin E)

Carotene samples are natural commercial products often used in food manufacturing and processing as colorants. Besides the unsaturated hydrocarbon-type carotenes which are a dietary source of vitamin A, the samples also contain oxygenated derivatives (xanthophylls) and tocopherols (vitamin E).

Analysis of carotenoids and tocopherols is normally performed by HPLC.[29–31] A commercial carotene sample was analysed by CEC on a 25 cm column using acetonitrile/tetrahydrofuran in ratio 95/5 to which 50 mM ammonium acetate was added. This is the second application of non-aqueous CEC. The sample dissolved in tetrahydrofuran was analysed with a voltage of

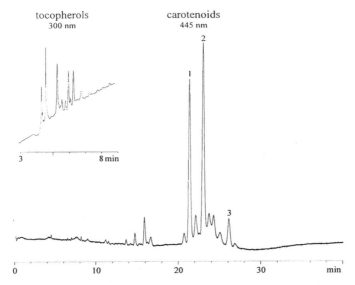

**Figure 8.14** *CEC analysis of a carotene sample. Column, 25 cm × 0.1 mm i.d., FSOT, Hypersil ODS 3 μm. Mobile phase, 50 mM NH₄Ac in THF AcCN (5/95). Detection UV at 445 nm except 0–10 min 300 nm. T' 20 °C; Voltage, 30 kV; injection 10 kV during 0.2 s. Peaks: 1, all-*trans*-α-carotene; 2, all-*trans*-β-carotene; 3, lycopene*

30 kV at 20 °C. Injection was performed electrokinetically at 10 kV during 0.2 s. As shown in Figure 8.14 the carotenoids elute after the tocopherols (insert). The carotenoids were detected at 445 nm and the tocopherols, eluting between 4 and 9 minutes, were detected at 300 nm. For this particular sample CEC provided a much more detailed picture than other separation methods, *e.g.* LC and SFC.

## CEC Analysis of a Fraction of Jojoba

Jojoba (*Simmondsia chinensis*) yields up to 60% oil, which has a substantial cosmetic value, and is rich in proteins (35%). The jojoba oil is composed of esters combining a fatty acid with a long carbon chain ($C_{18} - C_{20} - C_{22}$) and a double bond with a fatty alcohol ($C_{20} - C_{22} - C_{24}$) also with a double bond. The meal contains about 5% simmondsin and simmondsin-ferulates which are compounds with food intake inhibiting activity.[32–34] We recently isolated a fraction from the jojoba meal with interesting characteristics and submitted the fraction to several chromatographic techniques. Again CEC gave us by far the best separation (Figure 8.15) compared to LC and SFC. The mobile phase was 50 mM MES pH 6/MeOH (10/90). The analysis was performed at a voltage of 30 kV at 20 °C using UV detection at 280 nm. Injection was performed electrokinetically at 10 kV during 6 s.

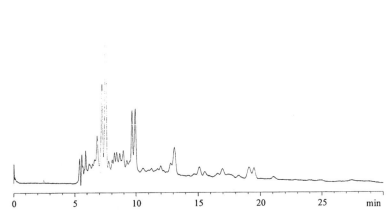

**Figure 8.15**    *CEC analysis of a jojoba meal fraction. Column, 25 cm × 0.1 mm i.d., FSOT,*
*Hypersil ODS 3 μm. Mobile phase, 50 mM MES pH 6/MeOH (10/90).*
*Detection UV at 280 nm. T' 20 °C; Voltage, 30 kV; injection, 10 kV during*
*6 s*

## 6  Conclusion

The various applications shown in this chapter emphasise the greater potential
of CEC. Unsurpassed efficiencies were obtained for samples normally analysed
by LC. Moreover the applications selected could not be carried out by CE and
micellar electrokinetic chromatography (MEKC) because of the lack of solubility of the analytes in aqueous media. Another advantage of CEC over CE and
MEKC is its MS compatibility, although CEC-MS is still in its infancy and an
optimum interface has to be developed. We hope that this chapter will
contribute to the further development of CEC and its acceptance. After all, . . .
was capillary GC not invented in 1960?

## References

1. M.M. Dittmann, K. Wienand, F. Bek, and G.P. Rozing, *LC-GC Int.*, 1995, **13**, 802.
2. W.W. Christie, 'Gas Chromatography and Lipids. A Practical Guide', The Oily
   Press, Ayr, 1989.
3. E. Geeraert and P. Sandra, *J. Am. Oil Chem. Soc.*, 1987, **64**, 100.
4. J.A. Garcia Reguerio, I. Diaz, F. David, and P. Sandra, *J. High Resolut. Chromatogr.*, 1998, **17**, 180.
5. W.W. Christie, 'High Performance Liquid Chromatography and Lipids. A Practical
   Guide,' Pergamon Press, Oxford, 1987.
6. E. Frede, *Chromatographia*, 1986, **21**, 29.
7. K. Aitztmüller and M. Grönheim, *J. High Resolut. Chromatogr.*, 1992, **15**, 219.
8. W.W. Christie, *J. High Resolut. Chromatogr.*, 1987, **10**, 148.
9. W.W. Christie, *Prog. Lipid Res.*, 1994, **33**, 9.

10. P. Laasko, in 'Advances in Lipid Methodology', ed. W.W. Christie, The Oily Press, Dundee, 1992, 82.
11. P. Sandra and F. David in 'Supercritical Fluid Technology in Oil and Lipid Chemistry', ed. J.W. King and G.R. List, AOCS Press, Champaign, IL, 1996, 321.
12. M. Demirbücker and L.S. Blomberg, *J. Chromatogr.*, 1991, **550**, 765.
13. V. Ferraz and P. Sandra, Proceedings of the 16th International Symposium on Capillary Chromatography, Riva Del Garda, Italy, 1994, 1544.
14. V. Ferraz, Ph.D. Dissertation, University of Gent, 1995.
15. P. Sandra, A. Dermaux, V. Ferraz, M.M. Dittmann, and G.P. Rozing, *J. Micro. Sep.*, 1997, **9**, 409.
16. G. Ross, M.M. Dittmann, F. Bek, and G.P. Rozing, *Int. Lab.*, 1996, **5**, 10A.
17. A. Dermaux, M. Ksir, K.F. Zarrouck, and P. Sandra, *J. High Resolut. Chromatogr.*, 1998, **21**, 545.
18. J. Dyerberg, H. Bang, and O. Aagaard, *Lancet*, 1980, **1**, 199.
19. F. Cambien, A. Jacquesson, and J.L. Richard, *Am. J. Epidemiol.*, 1986, **124**, 624.
20. W.S. Harris, W.E. Connor, S.B. Inkeles, and D.R. Illingworth, *Metabolism*, 1984, **33**, 1016.
21. A. Dermaux, A. Medvedovici, M. Ksir, E. Van Hove, M. Talbi, and P. Sandra, *J. High Resolut. Chromatogr.*, in press.
22. A. Medvedovici, M. Ksir, F. David, M. Talbi, and P. Sandra, *Chromatographia*, in press.
23. F. Cambien, A. Jacquesson, and J.L. Richard, *Am. J. Epidemiol.*, 1986, **124**, 624.
24. W.S. Harris, W.E. Connor, S.B. Inkeles, and D.R. Illingworth, *Metabolism*, 1984, **33**, 1016.
25. A. Medvedovici, F. David, and P. Sandra, *Chromatographia*, 1997, **44**, 37.
26. P. Sandra, A. Medvedovici, A. Kot, and F. David in 'SFC with Packed columns', ed. K. Anton and C. Berger, Marcel Dekker, New York, 1997.
27. A. Dermaux, V. Ferraz, and P. Sandra, *Electrophoresis*, 1999, **20**, 74.
28. R.C. Correa, V. Ferraz, K. Cerne, and P. Sandra, Proceedings of the 20th International Symposium on Capillary Chromatography, Riva Del Garda, Italy, 1998.
29. A. Ben-Amotz, *J. Liq. Chromatogr.*, 1995, **18**, 2813.
30. K.J. Scott, *Food Chem.*, 1992, **45**, 357.
31. W.T. Wahyuni and K. Jinno, *J. Chromatogr.*, 1998, **448**, 398.
32. C.A. Ellinger, A.C. Waiss, and R.E. Lundin, *J. Chem. Soc., Perkin Trans.*, 1973, 2209.
33. C.A. Ellinger, A.C. Waiss, and R.E. Lundin, *J. Org. Chem.*, 1974, **39**, 2930.
34. M.M. Cokelaere, H.D. Dangreau, S. Arnouts, R. Kühn, and E.M.-P. Decuypere, *J. Agric. Food Chem.*, 1992, **40**, 1839.

# Subject Index